LOOKING
FOR EARTHS

Looking for
Earth

The Race to Find
New Solar Systems

Alan Boss

John Wiley & Sons, Inc.
New York • Chichester • Weinheim • Brisbane • Singapore • Toronto

Copyright © 1998 by Alan Boss. All rights reserved.
Published by John Wiley & Sons, Inc.
Published simultaneously in Canada.

This publication is designed to provide accurate and authoritative information in regard to the subject matter covered. It is sold with the understanding that the publisher is not engaged in rendering professional services. If professional advice or other expert assistance is required, the services of a competent professional person should be sought.

Library of Congress Cataloging-in-Publication Data:

Boss, Alan
    Looking for earths : the race to find new solar systems / Alan
Boss.
        p.   cm.
    Includes index.
    ISBN 0-471-18421-7 (cloth : alk. paper)
    1. Extrasolar planets.   I. Title.
QB820.B68   1998
523—dc21   97-49033

Printed in the United States of America

10 9 8 7 6 5 4 3 2 1

To Catherine, Margaret, and Nicholas,
the centers of my universe

ACKNOWLEDGMENTS

The writing of this book would not have been possible without the knowledge I have gained from my colleagues, many of whom played active roles in the story described in *Looking for Earths*. Their names can be found throughout the text and the illustrations—I thank them all for their dogged efforts in pursuit of one of the most exciting goals in all of science.

My Carnegie Institution of Washington colleagues Robert Hazen and Frank Press, agent Gabriele Pantucci, and editor Emily Loose each played a key part in helping to turn a vague idea into a first book. Shaun Hardy and Merri Wolf operated a fine research library, in which facts can be tracked down in a few moments. Michael Acierno and Sandy Keiser kept the Internet accessible and the e-mail flowing. My Department of Terrestrial Magnetism colleagues, especially Sean Solomon, Louis Brown, John Graham, Vera Rubin, François Schweizer, and George Wetherill, provided a supportive environment.

Perhaps the largest debt I owe is to the people of the United States of America, whose financial support of NASA and the National Science Foundation has in large part enabled the scientific discoveries described in this book. On a more personal level, NASA provided the research assistantship that supported me for five years in graduate school, NASA supported my postdoctoral fellowship at the Ames Research Center, and NASA partially supports my ongoing research at the Carnegie Institution of Washington, as does the National Science Foundation. Stanton Peale, David Black, and George Wetherill, respectively, offered a position to me during these three critical phases of a scientist's career. I give them each my most profound thanks.

Alan Boss
Washington, D.C.
February 1998

Contents

PREFACE

An unprecedented event in the history of humanity is under way. For centuries, if not millennia, human beings have wondered about life on other worlds—are there planets elsewhere in the universe where creatures similar to *Homo sapiens* live and dream? Or are we alone in the unfathomable vastness of the universe? We are about to learn the answer to this anthropocentric illusion-shattering question.

Scientists have long guessed that planets like Earth should be common around the myriad of other stars in our galaxy, and that life surely has arisen on many other worlds. The commonly held assumption that we are not alone in the universe forms the basis for nearly all the science fiction books of this century, for blockbuster movies like the *Star Wars* trilogy and *Men in Black,* and for many popular television shows, from *My Favorite Martian* to the *Star Trek* phenomenon to *3rd Rock from the Sun.* But hard evidence for other solar systems, much less other advanced beings, has been nonexistent or even erroneous—the history of searches for new planets is one of unmitigated disaster.

Now all of that has changed forever. After laboring quietly for years in mountain-top observatories around the globe, several teams of astronomers have at last trapped unseeable wraiths that eluded capture for over half a century. As a result of their persistent efforts, we now have convincing evidence that *planets really do exist around other stars.*

A new era has just begun, an era in which we will discover many planetary systems circling stars in our neighborhood of the galaxy, systems containing Earth-like planets capable of supporting life. This book reveals the tortuous past that led to this epochal moment, as seen from the perspective of a scientist actively involved in the worldwide effort. Beginning with the first widely heralded discoveries of planets outside the Solar System decades ago, *Looking for Earths* tells the story of cyclical hope and despair as astronomers searched for irrefutable evidence of other planets, leading to the surprising triumphs of 1995–1996 and the dazzling prospects for discovering an Earth-like planet in our lifetimes.

PROLOGUE

The world was created on 22nd October, 4004 BC at 6 o'clock in the evening.

—James Ussher Archbishop of Armagh (1581–1656)

Human beings inevitably wonder about their past and future, where they came from, where they are going. Human societies argue about the possible answers to these eternal questions, sustaining the debate over time spans longer than that of a single human being, passing along the best ideas to successive generations by word of mouth or written records. The answers available at any given time will depend on the collective knowledge accumulated or acquired by the society. Primitive societies necessarily answer these questions in a simplistic fashion because they are limited in their knowledge of the natural world, which leads to creation myths in which a Great Fish, or worlds balanced on the back of an immense turtle, figure prominently.

With the hindsight of thousands of years of effort to understand the natural world, modern human beings have little use for the creation myths of primitive societies. Using our considerable understanding of the physical and biological nature of the universe, the dreamers among us can instead concoct a modern creation myth that reads more like science fiction than a biblical account of the creation of Heaven and Earth. The advance of science may be viewed as a human endeavor whose ultimate product is a scientifically based creation myth.

Ancient humans noticed that certain points of light in the night sky move with respect to other, apparently stationary points of light. The moving points of lights were called *planets,* a word derived from the Greek word for wanderer. The planets were thus distinguished from the *stars,* which seemed to be qualitatively different due to their fixed positions in the sky and vastly greater numbers. Only five planets were known in antiquity—Mercury, Venus, Mars, Jupiter, and Saturn—in

1

addition to the Sun and Moon, which also move past the fixed stars in the sky. These seven celestial wanderers were believed to be of such importance to humans that the seven days of the week were named after them.

The precise motions of the planets are so complicated, however, at times changing directions and making loops in the sky, that it took several millenia for human beings to figure out what in the world was going on.

In the sixth century B.C., the Greek mathematician Pythagoras tried to explain planetary motions by postulating the existence of a series of seven concentric, spinning spheres, each of which carried on its surface a planet, the Sun, or the Moon. Pythagoras' spheres were centered on the Earth, which was spherical itself, but fixed and unmoving. The notion that the Earth might move as well seems to have originated with Philolaus, a fifth century B.C. follower of Pythagoras. Philolaus suggested that the center of the universe was an unseen "central fire," about which the Earth, Sun, Moon, and planets revolved. These suggestions were based purely on imagination, though, not on scientific measurements.

Accurate measurements of the positions of the planets were first recorded by Hipparchus, a Greek astronomer from the second century B.C. Four centuries later, Hipparchus' carefully archived data allowed the Greek Ptolemy to propose an explanation of planetary motion that became widely accepted for the next thousand years. Ptolemy proposed that each planet moved about the Earth not on a sphere, but around the edge of a circle, the center of which itself moved around the edge of a second, larger circle. Like a light mounted on the leg of a person riding a bicycle at night, Ptolemy's epicyclic scheme produced motions that resembled the looping paths of the planets across the sky.

Ptolemy's epicyclic theory was accepted as dogma during the Middle Ages, and it was not until 1530 that Polish mathematician and astronomer Nicolas Copernicus finally showed that the motions of the planets and the Sun could be most simply explained if the Earth and planets moved on circles around the Sun, rather than the planets and Sun circling around the Earth. Copernicus showed that in his scheme, Mercury was the closest planet to the Sun, followed in order by Venus, Earth, Mars, Jupiter, and Saturn. Furthermore, the farther away each planet was from the Sun, the slower each planet moved. With these ideas, Copernicus was able to explain the looping motions of the planets in the sky. Copernicus also insisted that the daily rising and setting of the Sun and Moon are caused by the rotation of the Earth.

The breathtaking clarity of thought exhibited by Copernicus took some time to be accepted and built on. Danish astronomer Tycho Brahe made precise measurements of the positions of stars and planets in the late 1500s, but could not bring himself to believe in Copernicus's heliocentric universe. In 1609, German astronomer Johannes Kepler made the next great leap in understanding. In spite of his start as Brahe's assistant, Kepler improved the predictive ability of the heliocentic hypothesis by showing that the orbit of the planet Mars is not a circle, as Copernicus thought, but an ellipse (an oval, or flattened hoop), with the Sun near the center of the ellipse. Kepler also discovered a simple mathematical law that relates a planet's distance from the Sun to the time it takes to orbit the Sun.

If the Earth was not the center of the universe, but just one of several planets orbiting the Sun, then the Earth could hardly be the unique world enshrined in Ptolemy's Earth-centered universe. In addition, the Solar System itself might not be unique. The Catholic monk Giordano Bruno bravely asserted that the multitudinous stars in the sky are suns just like our own, and in 1600 he was burned at the stake in Rome for this heresy. The Catholic Church was not about to admit the possibility of life on other worlds, however indirectly. A few years later, in 1616, the Church solemnly declared the Copernican doctrine of a Solar System with its planets revolving around the Sun to be false, but it was too late: Copernicus's ideas could not be squelched by a religious edict.

In 1755, the German philosopher Immanuel Kant suggested that the Sun and planets of our Solar System had a common origin; this strengthened the presumed association between stars and planets. By 1786, the German English astronomer William Herschel had discovered a thousand fuzzy patches of "shining fluid" in the sky, termed nebulae, very unlike the sharp images of the pointlike stars and planets. The French mathematician Pierre Simon de Laplace speculated a decade later that the Sun and Solar System formed from one of Herschel's fuzzy nebulae.

Kant's and Laplace's reasoning was based on the knowledge that the planets orbit the Sun in a single plane, on nearly circular orbits, and in the same direction, which implies that the planets could have formed out of a single rotating disk of material with the Sun at its center. Given the apparently gaseous nature of Herschel's nebulae and of the Sun, it was natural to assume that the Sun and the disk that formed the planets had a shared origin. The Kant-Laplace hypothesis of coupled star and planet formation seemed reasonable, along with its implication that planetary systems may be as abundant as the stars themselves.

Herschel also accomplished the discovery of the first new planet in our Solar System (that is, of a planet unknown to ancient human beings). Herschel found a seventh planet on March 13, 1781, more or less by accident—he had no particular clue that another planet was hiding out there beyond Saturn, and at first thought he had just found a comet without a tail. Herschel wanted to name his spectacular discovery *Georgium Sidus,* after George III, king of his adopted country of England. In the face of opposition from non-English astronomers, however, he had to settle for the name Uranus, a word derived from the Greek word for heaven. King George appreciated Herschel's thwarted gesture anyway and rewarded Herschel by appointing him the royal astronomer, with a modest salary of 200 pounds per year, sufficient to allow Herschel to stop giving music lessons and to devote himself full time to astronomy.

On the nights of September 23–24, 1847, the German astronomer Johann Galle won a heated race with English astronomers to find a new planet whose existence had been predicted independently by two theorists several years before, the Englishman John C. Adams and the Frenchman Urbain Le Verrier. This eighth planet, circling even farther from the Sun than Uranus, was soon thereafter named Neptune.

Neptune's existence, and indeed its location in the sky, had been predictable because the gravitational force of Neptune exerts a continual pull on Uranus. Uranus refused to follow the orbital pattern that it would have followed if something fairly big was not out there continually yanking on Uranus and changing its course. Galle found Neptune right where Adams's and Le Verrier's calculations said it should be. Building on the solid foundations established by Copernicus and Kepler, Adams and Le Verrier achieved a prediction that was unrivaled in astronomy.

Herschel had shown in 1787 that the stars are not distributed at random in the sky, but lie in a gigantic disk, called the Milky Way galaxy. Only in the twentieth century was it proven conclusively that Herschel's fuzzy nebulae were not embryotic stars but distant galaxies, immense collections of stars just like the Milky Way. For every one of the hundreds of billions of stars in the Milky Way galaxy, there are many distant galaxies, each containing hundreds of billions of stars. The galactic nature of Hershel's nebulae meant that the known number of stars in the universe had increased almost immeasurably.

Possible havens for habitable planets outside the Solar System had become effectively infinite in number. Yet for nearly the entire twentieth century, it was not certain if there were any planets around even the closest stars in our galaxy. For all we knew, the Solar System might be as unique as Pythagoras and Ptolemy imagined it to be.

1

BARNARD'S REMARKABLE STAR

Do there exist many worlds, or is there but a single world? This is one of the most noble and exalted questions in the study of Nature.

—Albertus Magnus (thirteenth century)

APRIL 18, 1963: Peter van de Kamp was very pleased indeed. After decades of intensive work, the Swarthmore College astronomer succeeded at finding something incredibly elusive, something that human beings had wondered about for hundreds if not thousands of years, but could never be sure existed.

In a speech delivered to the nation less than two years before, President John F. Kennedy committed the United States to a space race with the Soviet Union, a race to land the first human beings on the Moon and to return them safely to Earth. Kennedy's visionary goal was no less than to extend the human presence to another celestial body, for the first time in the history of our planet. The space race to the Moon became a defining theme for the entire decade of the 1960s, the decade when humanity first escaped the confines of planet Earth.

But van de Kamp leapfrogged completely over Kennedy's vision—he firmly believed that he had found evidence for the existence of *a planet orbiting another nearby star*. Other astronomers had also sought evidence for such a new planet, but van de Kamp beat them to the prize. While the United States was preparing to take the first halting steps away from planet Earth, van de Kamp had already opened up an extraordinary vista, the possibility of completely new planetary systems to explore. The outcome of the lunar space race between the United States and the Soviet Union may have been uncertain in April 1963, but

5

FIGURE 1. Peter van de Kamp, professor of astronomy at Swarthmore College and Director of the Sproul Observatory, about 1958. For several decades, van de Kamp was the world's leader in searching for extrasolar planets. (Courtesy of Swarthmore College.)

it seemed clear that van de Kamp had won the race to find the first extrasolar planet, the first planet moving about a star other than our Sun.

The discovery did not come easily. Planets moving around other stars are nearly impossible to find. The visible light emitted by a planet like Earth or Jupiter is about one billionth that emitted by the Sun. A planet's clouds, oceans, ice sheets, and rocky terrain can only reflect the visible light that comes from the planet's star, rather than create their own visible light, like the Sun and other stars do. Because of their small sizes and considerable distances from their stars, planets reflect only a tiny portion of their star's light.

In comparison, stars are profligate sources of visible light. Because of the ability of stars to turn hydrogen into helium while liberating copious amounts of nuclear energy, stars can shine at prodigous rates for billions of years. Planets are denied this impressive ability and so are fated to remain nearly invisible from afar. Trying to see an extrasolar planet right next to its star is akin to trying to see a tiny mirror being held by a person standing next to a powerful carbon-arc searchlight that is pointed right at you—you can't see the light from the mirror because you are blinded by the far brighter light from the searchlight itself. Human beings could stare at the sky until their eyes popped without ever seeing a planet belonging to another star. The same was true of astronomers using the best telescopes and photographic film.

Van de Kamp knew he had no chance for such a direct detection of an extrasolar planet, so he used a more subtle and indirect means. Since the laws of nature are believed to be universal, any extrasolar planets must orbit their stars in the same way that the planets of our Solar System orbit the Sun, and therein lies a clever means to find them. Van de Kamp would look for the excrutiatingly small wiggles in the location of a star that would signal the presence of an unseen planet moving on an orbit around the star or, more correctly, around the center of the combined star and planet system.

For a star with a single planet, both the star and the planet move on nearly circular orbits about the center of mass of the two bodies. The center of mass is an imaginary point on a line between the star and the planet where the masses of the two bodies balance each other. On a playground teeter-totter carrying two children, the center of mass falls at the supporting bar at the middle of the teeter-totter. If an adult and a child try to balance on opposite ends of a teeter-totter, the adult must sit much closer to the central bar in order to balance out the lighter weight of the child. The center of mass is then much closer to the adult than to the child.

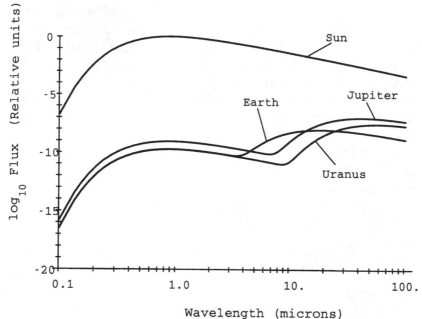

FIGURE 2. Discovering extrasolar planets by photography is difficult because planets are much fainter than their stars. Each unit on the (log) scale on the left represents a factor of 10 in brightness. At visual and near-infrared wavelengths (around 1 micron), the solar system's planets shine by reflected sunlight and are about a billion times fainter than the Sun. At mid-infrared wavelengths (about 10 microns), the planets give off thermal energy, leaving them still a million times fainter than the Sun. (Courtesy of Robert A. Brown, Space Telescope Science Institute.)

In the case of the Solar System, the largest planet, Jupiter, has a mass one thousand times less than that of the Sun. Because of this great imbalance, if Jupiter and the Sun both tried to balance on a teeter-totter, the Sun would have to sit one thousand times closer to the center than Jupiter would. In fact, in the case of Jupiter and the Sun, the center of mass lies at the outer surface of the Sun. As Jupiter orbits around the center of mass, the Sun moves in a circle with a radius roughly equal to the Sun's radius of 432,300 miles, a radius about one hundred times larger than that of the Earth. Jupiter's orbit is a thousand times larger (about five times the Earth's average distance from the Sun of 93 million miles), so Jupiter has to travel one thousand times faster in order to keep up with the slow-moving Sun.

The other planets in the Solar System have a similar effect on the Sun. The effect of the next largest planet, Saturn, is comparable to that

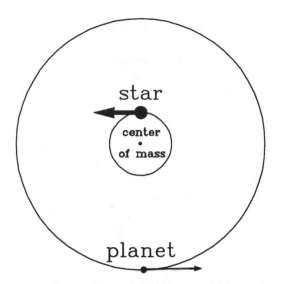

FIGURE 3. A star and its planet orbit around their common center of mass, the balance point for a star and planet on a teeter-totter. (Courtesy of Alan Boss.)

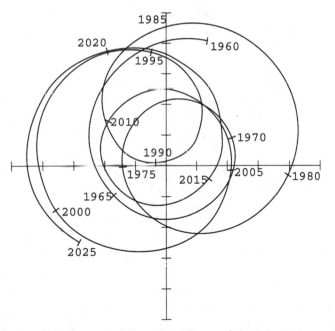

FIGURE 4. Motion of the Sun around the center of mass of the Solar System over a 65-year period. The Sun moves back and forth over a distance equal to its diameter. If seen from a distance of 33 light years, the tick marks on the axis would represent 0.2 milliarc-seconds of angular displacement. (Courtesy of Robert A. Brown, Space Telescope Science Institute. Based on JPL's DE200 planetary ephemeris.)

of Jupiter, because even though Saturn is three times less massive as Jupiter, it sits twice as far out on the teeter-totter and so can move the Sun around almost as much. Orbiting planets thus force their central star into a periodic dance, with rhythms and motions that can be used to infer the presence of otherwise unseen and perhaps unseeable objects.

This tiny stellar pirouette is exactly what Peter van de Kamp sought to discover. The effect is devilishly small, amounting to measuring the painfully slow motion of a star across a distance equal to its diameter during a time period perhaps decades in duration. The distance traversed by a nearby star dancing with a Jupiter-like planet is about the same as the apparent size of a dime viewed from a thousand miles away. With perseverance and very accurate astronomical measurements, dubbed astrometric measurements, van de Kamp believed he could perform the trick.

Though he was a master of the astrometric technique, van de Kamp did not invent this approach to finding unseen companions to stars. In 1844, the German astronomer Friedrich Wilhelm Bessel first found that the bright star Sirius underwent such a dance about an unseen companion. Sirius, in fact, is the brightest star in the sky, shining with a luminosity 26 times greater than that of our Sun, yet at a distance from the Sun of only 8.6 light years, making Sirius one of the Sun's closest stellar neighbors. (A light year is the distance that light travels in one year, or about 6 trillion miles. In just one second, light travels 186,000 miles, a bit less than the distance from the Earth to the Moon. A light year is a convenient unit for expressing distances to nearby stars, because the typical distance between adjacent stars in our galaxy is a few light years. The Moon is just 1.3 light seconds away from Earth.)

This first hidden stellar companion was not actually seen until 1862, when the American telescope maker Alvan G. Clark found Sirius' dark companion, a faint star giving off less than a thousandth the light emitted by its overachieving sibling. The bright star Sirius thereafter was known as Sirius A, while its faint companion was simply called Sirius B. Clark's feat was made possible by the fact that he had just constructed the largest telescopic refracting lens ever built, a polished glass disk with a diameter of a whopping 18 inches, considered monstrously huge in the Civil War year of 1862. The observations of Sirius were intended to be the ultimate test of the quality of Clark's lens, and the discovery of Sirius' companion made for a spectacular testimonial. Alvan Clark's Washington-based business of building telescopes prospered for quite some time thereafter, as orders for successively larger lenses arrived from observatories around the world.

By the time that van de Kamp arrived at Philadelphia's Swarthmore College in 1937, the astrometric method of finding unseen companions to stars was well established but not yet fully exploited. Van de Kamp used his new position as Director of Swarthmore's Sproul Observatory to begin in 1938 a long-term program to search for low-mass companions to stars. The search was primarily intended to find unseen stars, but it might be able to uncover a planet or two if a suitably close target star was scrutinized.

The telescope to be used for this program was Sproul Observatory's 24-inch refractor, only slightly larger in diameter than the refractor that Clark had used to see Sirius B (in refracting telescopes, light rays are bent or refracted into focus by passing through glass lenses, while in reflectors, the light is focused by bouncing off a mirrored surface). The search for unseen companions quickly became one of the primary objectives of the entire Sproul Observatory effort. While the goal was to find unseen stars like Sirius B, van de Kamp hoped that an exacting search would turn up evidence for a far grander prize.

The astrometric method of finding planets depends on the distance from Earth to the star under investigation. The small wiggle induced by an unseen planet becomes twice as hard to see in a star that is twice as far away, so the stars closest to the Sun are the best places to search for astrometric evidence of extrasolar planets. The closest star of all is Proxima Centauri (4.2 light years away), slightly closer to the Sun than the two stars of the Alpha Centauri system (4.4 light years). This group of three stars forms a triple system: The double stars in Alpha Centauri orbit about their common center of mass, while Proxima Centauri and Alpha Centauri (considered as a single object) orbit, in turn, about their mutual center of mass, making for a mind-churning three-body dance. After these three stars, the next closest star known is Barnard's star, at a distance of 6.0 light years.

Edward Emerson Barnard was an American photographer and amateur astronomer whose uncanny success at finding new comets led him to a professional career in astronomy at the Lick Observatory in California and then the University of Chicago's Yerkes Observatory. Comets are Manhattan-sized clumps of ice and dust that orbit the Sun in much the same way that the planets do, though on elongated orbits that often take them very far away from the Sun. Distant comets are faint and hard to find, yet still are tens of thousands of times closer to the Sun than the nearest stars. Because they are so close to the Sun, the apparent motion in the sky caused by their orbits around the Sun is quite large—they can be seen to move significant distances with respect to the fixed stars in just a night or two of observing. For this reason, comets can be found by comparing two photographs of a portion of the sky taken at two different times—if a speck of light appears to have shifted its location between the two photographs, there is a good chance that a new comet has been found.

Barnard took a photograph of the stars in the constellation Ophiuchus in August 1894, at Lick Observatory, and then another photograph of the same region early in 1916 at Yerkes Observatory. Barnard was amazed by what he saw when he compared the two photographic plates—one faint speck had fairly jumped past its neighboring stars in the 22-year interval between the photographs. This new star exhibited the largest such motion across the sky, termed proper motion, ever found. The idea that some stars are not truly fixed had been established well before—the English comet finder Edmund Halley had discovered in 1718 that several stars move perceptibly across the sky, including Sirius. However, the motions of stars through space are nearly impossible to detect unless the stars are very close by. A very distant airplane seems to lumber across the sky, even though it is moving at a speed of hundreds of miles an hour. Therefore, any star with a large proper motion has to be close to the Sun.

Barnard's star became the immediate focus of attention by other astronomers, who soon proved that the speedy star was indeed very close to the Sun, lying a mere 6.0 light years away, making it the closest star, with the exception of the Alpha Centauri triple system. Proxima Centauri had just been discovered the year before, in 1915, so the solar neighborhood suddenly had become considerably more crowded than it used to be.

Van de Kamp included Barnard's star in the list of nearby stars to be studied intensively in the long-term program at the Sproul Observatory—such a nearby star was a tempting candidate. The fact that the mass of Barnard's star is about seven times less than that of our Sun further improved the chances for detecting an unseen planet, because a lower-mass star is more easily thrown around by a planet of a given mass. In fact, astronomers at the Sproul Observatory had been observing Barnard's star from the beginning. Photographic plates (literally plates of glass, to give stability and permanence) of Barnard's star had been taken at Sproul as far back as 1916, and these plates provided van de Kamp with something of a head start in the race to find an extrasolar planet.

Van de Kamp had to be a very patient man in order to pursue work that required decades of painstaking effort before the result was obtained. In fact, he had presented one analysis of the wobble of Barnard's star 19 years before his 1963 announcement. The 1944 announcement was delivered at a meeting of the American Philosophical Society, rather than the expected venue of the American Astronomical Society, because of the proximity of the Philosophical Society meetings, which are held in Philadelphia (wartime travel was difficult), and because of the immense prestige of the Philosophical Society for those who live in the Philadelphia area. Van de Kamp stated at the meeting that he had evidence for an unseen companion to Barnard's star with a mass about 6 percent that of the Sun, a very low-mass star indeed, but at 60 Jupiter masses, clearly not a planet. Van de Kamp proved to be much more cautious about making claims about extrasolar planets than some of his peers.

Several astronomers had made extraordinary claims the previous year. Dirk Reuyl and Erik Holmberg of the McCormick Observatory of the University of Virginia published a paper in January 1943 stating that they had detected the presence of an "invisible companion" lurking in a nearby double star called 70 Ophiuchi. The name 70 Ophiuchi derives from the astronomical tradition of systematically naming stars by Greek letters, numbers, and constellation name (constellations are groupings of stars on the sky). Thus 70 Ophiuchi is so named because it is the seventieth numbered star in a listing of the brightest stars in the constellation Ophiuchus, by chance the same constellation that contains Barnard's star.

Whereas the Sun is a single star, 70 Ophiuchi is a double star. A double (or binary) star consists of two stars orbiting about their common center of mass. As long as a planet orbits very close to one or the other star, it can orbit happily without running into either star or being kicked out of the group. A planet could also orbit happily for a long period of time if it traveled in a circle much larger in size than that of the binary stars' orbit, because when two stars are that close, they act on the planet much as if they were a single star made by adding the two stars together. Thus it is conceivable that planets might exist as part of binary star systems, as well as around solitary stars like our Sun.

Reuyl and Holmberg used data taken during the previous decade to show that, after adjusting the observations to account for the wobble of the binary stars about their center of mass, there remained a cyclical wobble that could be caused by a 10-Jupiter-mass planet orbiting one of the stars. They avoided calling their newfound object a planet, however.

One month later, in 1943, Danish American astronomer Kaj Aage Strand, of the Sproul Observatory, published evidence that he had found a "planet" 16 times as massive as Jupiter, using eight years of data. The "planet" orbited one of the stars in a binary system called 61 Cygni, located in the constellation Cygnus. In spite of the fact that his object was 16 times heavier than the largest planet in our Solar System, Strand argued that he had indeed shown that "planetary motion has been found outside the Solar System." Strand had also studied the 70 Ophiuchi system but had found no evidence for a third body there, contrary to the data of Reuyl and Holmberg.

Both claims received considerable attention from the press and from other astronomers. Questions were raised about what had really been found—a columnist in the magazine *Sky and Telescope* sagely warned readers that just because a companion is invisible and low in mass, it need not be a planet.

The American Astronomical Society met for its annual meeting in Cincinnati, Ohio in November of 1943 and held a special symposium on "Dwarf Stars and Planet-like Companions." Strand was scheduled to speak but did not attend; van de Kamp did attend and spoke on Strand's behalf. Strand's 61 Cygni result was presented by van de Kamp as evidence of an object of "intermediate mass." Van de Kamp used this designation because the mass of Strand's object fell into limbo between the masses of the largest known planet (Jupiter) and the least massive star known at the time (Krüger 60 B, with about one seventh the mass of the Sun).

Reuyl and Holmberg were not on the program for the symposium, and the McCormick Observatory claim for 70 Ophiuchi's third body was not mentioned in the meeting report. Given Strand's earlier negative result on 70 Ophiuchi, and van de Kamp's position as Strand's director, it is perhaps not surprising that the claim from the competition was not stressed. When an astronomer has a problem with another astronomer's work, a common method of handling the problem is to act as if the other work does not exist. Personal confrontations over competing claims can then be minimized, if not eliminated. If the competitor is not present in the audience, no one may even notice the omission. If challenged, astronomers can then claim that they failed to mention a competitor's results simply because they weren't given enough time for their presentation.

The influential British science journal *Nature,* founded in 1869, went along with the claims that "non-solar planets" had been discovered. Writing about the claims for 61 Cygni and 70 Ophiuchi, Dr. A. Hunter stated that the data "points very strongly to the presence of companions the mass of which is so small that they would be more accurately described as planets than as stars." Hunter was thus willing to describe the third body in 70 Ophiuchi as a planet, based on the belief that its mass was only 10 times greater than that of Jupiter, an interpretation that Reuyl and Holmberg hesitated to make, though they must surely have hoped it to be true. The belief that planets had been found in the 61 Cygni and 70 Ophiuchi systems thus gained some credence with *Nature*'s blessing.

The definition of what constitutes a planet is not clear cut. Van de Kamp was of the mind that an object with a mass 10 or more times greater than that of Jupiter was *not* a true planet, but simply a small-mass star that gave off very little light. Van de Kamp thus regarded the announcements of new planets in 1943 as very premature. In the succeeding years, other claims were made for the detection of unseen companions with masses on the order of 10 Jupiters. In many of these cases, van de Kamp did not believe anything had been detected at all; most likely, overly ambitious astronomers were just looking at noise. Van de Kamp published his assessment of the various claims for detections of extrasolar planets in 1956. He specifically criticized the claims for 70 Ophiuchi as being spurious, and he concluded that all the results so far were "essentially negative, or at best tentative."

By 1963, however, van de Kamp was ready to change his mind about his Barnard's star data and make his own announcement of the discovery of a planet. The measurements he had made in the years since 1944 had reduced his likely sources of error, allowing him to detect the presence of much lower-mass objects. By taking about one hundred photographs of Barnard's star each year, van de Kamp's Sproul Observatory had amassed by 1963 an archive of over 2,400 photographic images of Barnard's star and its neighbors, taken by 50 different astronomers with the Sproul 24-inch telescope.

The location of Barnard's star and neighboring stars had to be meticulously measured on each of the 2,400 plates, using a specialized machine, and the measurements were repeated by different people in order to remove the errors caused by different levels of skill. Positions had to be measured to an accuracy much finer than the tiny size of the fuzzy blur on the photographic plate that was Barnard's star. The plates may have been taken with a large telescope, but in the end they had to be examined with a microscope.

The intensive effort at Sproul reached a conclusion on April 18, 1963 at the American Astronomical Society meeting in Tucson, Arizona. Van de Kamp announced that Barnard's star was circled by an object with a mass only 60 percent greater than that of Jupiter—surely

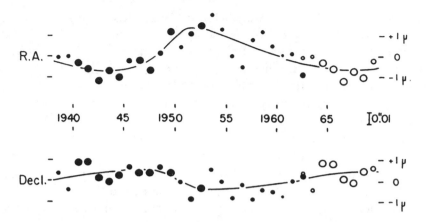

FIGURE 5. Van de Kamp's evidence for a planetary companion to Barnard's star. The shift of the position of Barnard's star in two directions on the plane of the sky is shown over the 24-year period inferred for the planet's orbit. The solid line is the best fit to the data points, open circles are earlier data points that have been carried forward in time and repeated, and the size of the points indicates the confidence level. (Reprinted, by permission, from P. van de Kamp, 1963, *Astronomical Journal,* volume 68, page 515. Copyright 1963 by the American Astronomical Society.)

this was a true planet! The 46 years of photographs used in the analysis implied that the planet orbited about its star once every 24 years, twice as long as Jupiter's orbital period of 12 years. The wobble of Barnard's star that was used to deduce this planetary orbit was a *hundred times smaller than the image of the star itself*—it was quite a feat to measure such a small shift in the center point of the blurred image of the star.

Van de Kamp was very pleased with his accomplishment, and he wrote a popular article for *Sky and Telescope* magazine that appeared several months before his results were formally published in the *Astronomical Journal,* a venerable compendium of American astronomical research. A long, long wait was over at last, and van de Kamp relished the attention he was sure to receive. He was, after all, a bit of a showman, prone to playing the piano at parties of astronomers until well into the morning hours, to wearing ascot scarves and a trademark Shetland Island woolen sweater, and to driving a succession of what were considered racy cars at the time, such as a red Pontiac convertible. Van de Kamp even served as the conductor of the Swarthmore College symphony orchestra for a decade. With the Barnard's star work, his decades of supervision of the Sproul astrometry program had finally reached fruition, and he was ready to reap his reward.

Van de Kamp's announcement received only a restrained welcome, however. The *New York Times* published a fourth-page story about the discovery on April 19, 1963, pointing out that this was the "third such body" found and lumping it together with the two previous claims for "planets" 10 times more massive than van de Kamp's. This attitude downplayed the importance of van de Kamp's apparent discovery of the first truly planet-sized body outside the Solar System. *Nature* published a paragraph about the discovery in its November 30 issue but avoided the use of the word *planet* in referring to Barnard's star's new companion. *Science,* the official journal of the American Association for the Advancement of Science, fully as prestigious a journal as *Nature,* did not feature this American discovery at all.

Seven months after van de Kamp's announcement at the Tucson meeting, President Kennedy was assassinated, though his dream of putting a man on the Moon lived on in the United States. The space race with the Soviet Union continued at a fever pitch, with the United States expending billions of dollars each year to try to reach the Moon first. Van de Kamp may have been discouraged by the world's lack of enthusiasm for his discovery of the first extrasolar planet, but he doggedly continued his life's work of finding dim companions to stars.

2

OTHER STARS, OTHER PLANETS

The orbital analysis leads, therefore, to a perturbing mass of only 1.6 times the mass of Jupiter. We shall interpret this result as a companion of Barnard's star, which therefore appears to be a planet.

—Peter van de Kamp (1901–1995), September 1963

Van de Kamp firmly believed he had succeeded in finding the first planet outside the Solar System. Jupiter is the most massive object in our Solar System besides the Sun, weighing 318 times more than the Earth does. By analogy with our Solar System, because of its mass van de Kamp believed that his new planet was similar to Jupiter, just a bit more obese.

Jupiter is called a gas giant planet, because most of its bulk is composed of hydrogen and helium, the two simplest and lightest elements. Hydrogen and helium are gases at the conditions prevailing at the surfaces of the Earth and Jupiter. Because hydrogen and helium are so light, they are used to inflate balloons and dirigibles (though hydrogen is also extremely flammable, as the passengers of the German airship *Hindenburg* learned when it burned explosively in 1937). The hydrogen in Jupiter's atmosphere is in the form of a molecule composed of two atoms of hydrogen. At high enough pressures, hydrogen molecules split up and form monatomic hydrogen, which conducts electricity like a metal. The deep interior of Jupiter is at such high temperatures and pressures that it may be a metallic liquid, but probably not a solid. In that case, Jupiter has no solid surface outside its core.

Because Jupiter is composed primarily of the two lightest elements, hydrogen and helium, in spite of its great mass Jupiter's average density is about the same as that of water, about one gram per cubic centimeter. The other gas giant planet in our Solar System is Saturn, with a mass 95 times that of Earth, composed again mostly of hydrogen and helium. However, Saturn is even less dense than Jupiter and would actually float if there was an ocean large enough to try the experiment.

Both Jupiter and Saturn are believed to contain cores at their centers composed mainly of solid rock and ice, buried deep within their overlying mantles of hydrogen and helium. If not for these rock and ice cores, the compositions of Jupiter and Saturn would be very similar to that of the Sun and most other stars, which consist of about 70 percent (by mass) hydrogen, 28 percent helium, and 2 percent elements heavier than helium. The heavier elements are the silicon, oxygen, iron, and other atoms that make up the bulk of the mass of terrestrial planets like Earth.

While van de Kamp could not tell exactly what sort of planet he had found orbiting Barnard's star, the fact that its mass was so close to that of Jupiter implied that he had bagged a gas giant planet. In addition, the 24-year period of the planet's orbit meant that the planet's distance from its star was very similar to that of Jupiter's distance from the Sun: 4.4 times the Earth-Sun distance, only slightly less than the value (5.2) for Jupiter. In terms of mass and orbital size, then, Barnard's star seemed to have a Jupiter clone.

There was one major difference, however—the planet orbiting Barnard's star was believed to be on a highly elliptical orbit, shaped like an oval four times longer in one direction than in the other. This orbit was unlike that of Jupiter, which is nearly a circle. Most of the other planets in the Solar System also have nearly circular orbits; the orbit of van de Kamp's planet was unlike that of any other known major planet. On the other hand, binary stars often have elliptical orbits, so the idea of a companion on an elliptical orbit did not necessarily seem out of the ordinary, given what was known at the time about how stars and planets form. The shape of the orbit would prove to be a critical issue in extrasolar planet searches.

If Barnard's star, one of the Sun's closest neighbors, could (just barely) be shown to be accompanied by a gas giant planet, then surely there must be many more gas giants orbiting the myriad of stars farther away from the Sun, stars too distant for the astrometric technique to be able to measure their tiny wobbles. Surely if we were only closer to these stars, van de Kamp could detect planets in their shadows as well. And if the analogy with our Solar System was correct, these stars would shelter entire systems of planets, Earths and Venuses, Neptunes and

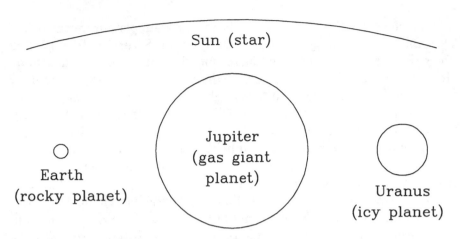

FIGURE 6. Relative sizes of the three types of planets in our Solar System compared to that of the Sun, a typical star. (Courtesy of Alan Boss.)

Uranuses. The most intoxicating implication of van de Kamp's discovery was that planetary systems were common in the universe—wherever we stare at the night sky, wherever we see stars, we are also looking at other solar systems.

Giordano Bruno had been burned at the stake just for hinting that planets orbited the stars, and now van de Kamp seemed to have proven Bruno's costly implication. Van de Kamp's discovery also gave great support to the Kant-Laplace hypothesis of coupled star and planet formation, proposed centuries before, only to be discarded at the opening of the twentieth century.

In 1900, a graduate student at the University of Chicago, Forest Ray Moulton, argued that the nebular hypothesis of Kant and Laplace should be ruled out because it would lead to the formation of a rapidly rotating Sun, contrary to reality. The Sun takes a leisurely month to complete a single rotation, versus a day or less for the Earth and most of the other planets.

Laplace had envisioned that a rotating cloud of gas and dust would contract into a progressively smaller and smaller object because of the inward pull of its gravitational attraction. Because of the effects of rotation, the contracting cloud would be flattened into a disk during this process, pulled outward by the same centrifugal force that a pizza maker uses to make a thin pizza crust out of a rapidly spinning handful of dough. As the primordial disk continued to contract and flatten, Laplace imagined that it would spit out successive rings of gas that would be left behind to condense later into the planets. The process required a central object spinning fast enough to spit out rings, and with no other way known to slow it down, Moulton argued that a Sun formed by this process should be left in a state of rapid rotation. The Sun's slow rotation then required that the nebular hypothesis be abandoned.

In place of the nebular hypothesis, Moulton and the chairman of his department at Chicago, the geologist Thomas C. Chamberlin, offered their own theory. Chamberlin and Moulton proposed in 1905 that the planets formed from gaseous filaments pulled out of the Sun by tidal forces during a near collision between the Sun and another star, now long gone. Tidal forces, which tend to pull a fluid body into a cigar shape, are caused by the gravitational pull of one body on another. For example, the Moon is largely responsible for the ocean tides on the Earth. The Moon's gravity causes oceanic water to form a cigar-shaped tidal bulge, centered on opposite hemispheres of the Earth, with one end pointing toward, and the other away from, the direction to the Moon. The height of the ocean then rises and falls roughly twice a day as a given Earth location rotates into and then out of each of the two tidal bulges.

Any two bodies that come close together will exert tidal forces on each other. Chamberlin and Moulton suggested that hot gases could be extruded from a star by a close encounter with another star. These gases would be left in orbit about the star and would eventually cool and condense to form a planetary system. Like Laplace over a century before, Chamberlin and Moulton were mistaken about the true nature of Herschel's nebulae, and suggested that what turned out to be two-arm spiral galaxies were examples of stars in the process of being pulled apart by the tidal force of other stars.

The tidal hypothesis had the merit of explaining why the planets all orbit in a single plane and in a single direction: A slender tidal filament slung around the Sun could explain both of these basic facts. The filament would be whipped around the Sun like a gigantic weed trimmer's plastic string. The tidal hypothesis could also explain how the planets could be formed out of a slowly rotating star like the Sun.

However, the tidal hypothesis carried with it the implication that planetary systems were rare, because stars essentially never pass close enough together for tidal forces to come into play. Tidal forces only become strong when two bodies are close to touching, yet stars are separated by distances that are, on average, about 40 million times too large. The chances of two stars undergoing a close encounter were thus infinitesimally small. If Chamberlin and Moulton were right, our Solar

System was likely to be the only planetary system in the entire Milky Way galaxy. We would truly be alone in our galaxy, at least, if not in the universe.

Other eminent scientists came to believe in Chamberlin and Moulton's tidal hypothesis—notably two British scientists, the applied mathematician James H. Jeans and the geophysicist Harold Jeffreys. The support of these two highly regarded scientists seemed to amount to a death knell for the hoary Kant-Laplace hypothesis. Given that no credible evidence for other planetary systems had been presented prior to the claims of 1943, the lonely universe implied by the tidal hypothesis seemed inescapable to Jeans when he developed his own variant of the idea in the depressing war year of 1917.

Despite its merits, the tidal hypothesis itself did not go unchallenged for long. In 1934, the American astronomer Henry Norris Russell soundly debunked the tidal theory. He pointed out that a filament of matter pulled out of the Sun would rapidly disperse into space because of the high temperature of the Sun's outer layers. Even the relatively cool outer layer of the Sun that we see has a temperature of about 5,500 degrees Centigrade, or 10,000 degrees Fahrenheit, and the deeper layers are much hotter yet. Solar material is so hot that it is termed a plasma, meaning that all molecules have been destroyed and most electrons have been stripped off the atoms. If freed from the Sun's gravity by tidal forces, the electrons and atoms of a million-degree plasma would zip off into outer space and never be seen again. About the last thing they would do is stick around and form a planetary system.

Russell's refutation of the tidal hypothesis left many scientists with the feeling that they now had no acceptable theory of how planets formed at all.

The claimed discovery in 1943 of "planets" around 70 Ophiuchi (by Reuyl and Holmberg) and around 61 Cygni (by Strand) partially cleared this mental logjam about planetary formation. These astrometric observations seemed to imply that planetary systems truly were abundant, so any theory that did not make copious planets was clearly in serious trouble. The tidal theory thus seemed to be doubly doomed.

Jeans (Sir James Jeans, as of 1928) was still not willing to give up on the tidal theory, and, writing to *Nature* from his retirement home in England in 1943, he stated he had already updated his tidal theory to allow planetary systems to be more prevalent. Jeans's 1942 *Nature* paper had increased the likelihood of forming tidal filaments by hypothesizing that the tidal interaction occurred when the Sun was much younger and perhaps 5,000 times larger in size than its present radius.

With such a vastly larger target to hit, the likelihood of a close encounter between a severely bloated early Sun and another star increases by a factor of 25 million, to a more respectable number that Jeans argued could account for the 70 Ophiuchi and 61 Cygni "planets."

By appealing to tidal interactions with an earlier, much expanded Sun, Jeans was inadvertently moving in the direction of the Kant-Laplace nebular hypothesis, which also invoked an early phase when the Sun was spread out over a much greater volume of space. But what really started astronomers thinking seriously about the nebula hypothesis again was the appearance in 1944 of a paper by the German physicist Carl-Friedrich von Weizsäcker, immediately after the publication of the 70 Ophiuchi and 61 Cygni claims. Remarkably, the fact that a horrendous world war was then raging did not seem to slow the pace of discovery or inhibit the free exchange of ideas between the United States and Germany.

Von Weizsäcker proposed that the nebular hypothesis could be salvaged by pointing out that the solar nebula undoubtedly was composed mostly of hydrogen and helium gas, like the Sun. The elements needed to form rocky planets would be a small fraction of the total mass of that portion of the nebula that made the terrestrial planets. Because only a few percent of the disk's mass would then be needed to form planets like the Earth, von Weizsäcker was confident that he could solve the Sun's rotation problem by proposing that when the unused hydrogen and helium gas escaped from the planet-forming disk, it must have carried off the unwanted rotational tendency (i.e., angular momentum) as well. The loss of angular momentum would occur naturally because of the rotation of the escaping gases themselves. The Sun could then be rescued from having to be in rapid rotation. While the means for effecting this miracle was not yet clear, von Weizsäcker had hit on a seductively attractive means for resurrecting the nebular hypothesis. If von Weizsäcker's inspired guess was correct, the universe was allowed to have planets aplenty once again.

Von Weizsäcker carried these ideas with him when he visited the University of Chicago in the late 1940s and thereby infected two influential scientists with the nebular bug. In spite of Chicago's being the home of Chamberlin and Moulton, the original proponents of the tidal filament hypothesis, Harold Urey and Gerard Kuiper listened to von Weizsäcker's ideas and became converts to the nebular hypothesis. Urey and Kuiper trained many of the next generation of scientists who would worry about planetary formation. The nebular hypothesis was ingrained in them from the beginning of their studies. The tidal theory was dead; it made sense once again to search for planets around other stars. The news of van de Kamp's 1963 discovery thus fell on receptive ears in the planetary science community.

Van de Kamp's discovery of the first extrasolar Jupiter put him in a class by himself. In terms of discovering new planets, though, he joined a small fraternity, a group with but a single living member at the time, an American named Clyde Tombaugh.

The ninth and last planet to be discovered in our Solar System was Pluto. Though the actual discovery by Tombaugh had been made on February 18, 1930, the director of Arizona's Lowell Observatory, V. M. Slipher, decided to wait until March 13 and issue a press release at the same time that astronomers worldwide would receive official word by telegraph. On the chosen day, the world learned that Lowell Observatory's Clyde Tombaugh had discovered Pluto, the most distant planet of all. The calculated press release worked—the discovery became the front page story of the *New York Times* on March 14, 1930 and caused a worldwide sensation.

Tombaugh ended up finding Pluto not through some inspired theory or serendipity, but simply through brute force. Tombaugh had been hired in January 1929 specifically to undertake a systematic search of the outer Solar System in hopes of finding a ninth planet. The founder of the Lowell Observatory, Percival Lowell, had believed that small discrepancies in the orbits of Uranus and Neptune required the presence of a ninth planet, which he called "Planet X." Lowell had searched for Planet X himself, but died in 1916 without achieving his goal.

Tombaugh was an amateur astronomer who had sent his sketches of the planets to the Lowell Observatory in hopes of getting expert advice on how to do a better job of using his homemade telescope. Instead of receiving advice, Tombaugh was hired to use the Lowell Observatory's newly installed 13-inch telescope to photograph the sky in the direction where Lowell had predicted Planet X should be. The Director himself searched the first several photographic plates for Lowell's Planet X, but he found nothing in the predicted location. The astonishingly successful story of the prediction and discovery of Neptune apparently was not going to be repeated with Planet X. Clearly disappointed and discouraged, Slipher turned over the entire search effort to Tombaugh.

Tombaugh began the search upon his arrival at Lowell, spending long winter nights in an unheated observatory at an elevation of 7,000 feet. He continued to take photographs of the band in the sky known to contain the other planets, and he compared photographs taken at different times to look for the tell-tale motion of some faint object fairly hurtling across the sky. Because of its nearness, the motion of an unseen planet would be large enough that Tombaugh could search for this motion with just his eyes. Though punishing work by most standards, Tombaugh's long hours of staring at pairs of photographic plates were more palatable to him than the Kansas wheat farmer's life that he fled in order to work at Lowell.

Tombaugh's perseverance paid off. Just one year after arriving at Lowell, Tombaugh found what he was looking for, and the Solar System had a new planet. The planet was duly named Pluto, for the mythological god of the underworld, but Pluto's first two letters also happened to be the same as the initials of Percival Lowell.

The orbit of Pluto turned out to be reasonably close to that of Lowell's hypothetical Planet X. However, because of its faintness, it was obvious that Pluto was nowhere near being massive enough to cause the variations in the Uranian and Neptunian orbits that Lowell had sought to explain with Planet X. Perhaps there was yet another planet still remaining to be discovered? Tombaugh was encouraged by Slipher to continue his successful search program. Unfortunately, the next 13 years of searching yielded no more planets. Tombaugh reluctantly concluded that Lowell's Planet X did not exist—the small orbital variations of Uranus and Neptune would require some other, more prosaic explanation.

Van de Kamp's perseverance had paid off too, though in an opposite fashion from Tombaugh's. Tombaugh had found his quarry the first year and then spent a fruitless decade looking for the next. Van de Kamp had labored for several decades before being rewarded with his first extrasolar planet.

Van de Kamp continued to monitor Barnard's star, which had shown evidence for the lowest mass object ever found orbiting another star, even if the astronomical community had conservatively shied away from calling it a planet. Van de Kamp was similarly cautious about his extrasolar planet claim. In a 1964 report to the International Astronomical Union, writing in his capacity as President of the Commission on Double Stars, van de Kamp tersely noted in his summary of the work of the last three years that he and his colleagues at Sproul were engaged in a "search for perturbations"; no mention was made of his epochal discovery of the previous year.

3

BARNARD'S STAR
CHANGES ITS COURSE

To measure is to know.

—Ernst Werner von Siemens (1816–1892)

APRIL 1–4, 1968: Five years later van de Kamp had amassed enough new observations of Barnard's star to redo his 1963 analysis, using measurements from what had grown to over 3,000 photographic plates spanning the half century from 1916 to 1967. He presented his new results in April at the American Astronomical Society meeting in Charlottesville, Virginia, home of the McCormick Observatory, where he had started his career in American astronomy in 1923. With the 500 new plates, van de Kamp's best estimate was that Barnard's star was circled by a planet 70 percent heavier than Jupiter, in an orbit 4.5 times larger than Earth's, values that were only slightly larger than his 1963 estimates. The unseen planet continued to resemble Jupiter, except for its peculiarly noncircular orbit, now with an even larger elongation than before. Van de Kamp had confirmed his own 1963 discovery.

While van de Kamp's confirmation did not spark much of a flurry of press reports beyond a mention in *Sky and Telescope,* the astronomical community finally began to warm to the concept that an extrasolar planet had actually been discovered. University of California, Los Angeles (UCLA) astronomer George Abell put van de Kamp's discovery into the 1969 revised version of his extremely popular astronomy textbook, stating that "this is the only object of planetary mass known beyond the Solar System." Prior to 1969, most textbook authors were cautious in asserting that any extrasolar planets had been discovered—in particular, the 10 to 16 Jupiter-mass companions claimed for 70 Ophiuchi and 61 Cygni, respectively, were not necessarily planets, if they were real at all. No other astronomers had come forward with strong evidence in support of the reality of the "planets" claimed for 70 Ophiuchi and 61 Cygni, and some astrometrists, including van de Kamp, were openly dismissive of them. The 1943 claims slipped quietly into the past without any further mention. By 1969, Barnard's star was about the only show in town.

The highly noncircular orbit derived for Barnard's star's companion bothered van de Kamp—it simply did not look like a planet's orbit should look, based on our Solar System. For another thing, a planet on an elongated orbit moves close to its star during part of its orbit, and then much farther away. Because of this periodic sweeping back and forth, a 1.7 Jupiter-mass planet would pass close to any Earth-like planets that might be trying to orbit safely between Barnard's star and the giant planet, circling around nervously like a jockey trying to referee a heavyweight fight. A giant planet on an oval orbit would quickly kick a measly Earth-mass planet out of the way without a second thought. This certainty did not bode well for the ancient dream of finding another planetary system capable of supporting life. Van de Kamp's paper, published in the March 1969 *Astronomical Journal*, did not mention this vexing aspect of the discovery.

APRIL 14, 1969: Van de Kamp had a brainstorm. He had come up with a completely different interpretation of his Barnard's star data that would solve the problem of the noncircular orbit. He presented the new analysis in a lecture at Duke University in North Carolina. The *New York Times* responded with an article about it the next day, though the result would not be published in the *Astronomical Journal* until August.

Van de Kamp realized that he could fit the observations of Barnard's star just as well with *two planets* on *circular orbits*. By putting in the tugs of two planets with circular orbits and periods of 12 and 26 years, van de Kamp found that he could do just as good a job of making Barnard's star dance about as he could with a single planet on an oval orbit. The inner planet had to have a mass 80 percent that of Jupiter on an orbit at 2.8 times the Earth-Sun distance, while the outer planet had a mass only 10 percent greater than Jupiter's and orbited at 4.7 times Earth's distance. Furthermore, van de Kamp found that he could explain the observations by having the two new planets orbit more or less in the same plane, as is the case for the planets in our Solar System.

The payoff was clear. By adopting this new solution, van de Kamp produced a planetary system that looked a lot more like our Solar System: a system with two massive, gas giant planets, moving on circular orbits in the same plane, and at distances comparable to Jupiter's distance from our Sun. The fact that the orbital radii were a bit smaller than in our Solar System could be attributed to the fact that Barnard's star has only 15 percent of the mass of our Sun, and a smaller star might be expected to have a more compact planetary system.

Furthermore, simply by reinterpreting his data van de Kamp had *doubled* the number of extrasolar planets he had found. That is the sort of bonus that usually only theoreticians get to enjoy—a new discovery without having to spend a single night at the telescope.

Others decided to play the game as well. One enterprising duo performed their own analysis of van de Kamp's 1969 data, using a popular statistical approach called the maximum entropy method, which maximizes the information that can be pulled out of a data set. They concluded that Barnard's star was circled by not one or two planets, but by *five* planets with masses ranging from 70 percent to 1.6 times the mass of Jupiter. They later found an error in their analysis and retracted two of their planets, but claimed there was still evidence for three planets around Barnard's star.

JULY 20, 1969: My family and I were celebrating my eighteenth birthday at our home in Florida. After the birthday cake was cut and devoured, we sat down together to watch a black and white television picture showing astronaut Neil Armstrong, who was at that moment stepping down off the ladder of the *Apollo 11* lunar module and onto the surface of the Moon. After eons of evolution, humankind had developed both the ability and the desire to transport themselves physically to other worlds. Though the Moon is by far our nearest neighbor in space, and hence the easiest to reach, the milestone achieved by the *Apollo 11* mission was a worldwide sensation and an unforgettable eighteenth birthday present.

The United States' National Aeronautics and Space Agency (NASA) was founded in 1958 as a direct reaction to the surprising success of the Soviet Union in launching the first artificial satellite of the Earth, *Sputnik I,* on October 4, 1957. The Soviet triumph should not have been the surprise it was—Soviet exhibits at scientific meetings held prior to the launch of *Sputnik I* had proclaimed the Soviet Union's plans to launch Earth satellites in the near future. By the rules of the then raging Cold War, the Soviet success with *Sputnik* dictated an American response, and the space race was on.

With the flight of *Apollo 11*, NASA met President Kennedy's 1961 challenge to land a man on the Moon and bring him back alive before the end of the decade. The United States decisively beat the Soviet Union to the grand prize of the space race.

Working largely in secrecy, the Soviets had accepted the challenge of the race to put a man on the Moon, but a sequence of failures of their launch vehicle put them so far behind the highly publicized American effort that they decided to give up. For propaganda purposes, the Soviets insisted that their goal instead was to return the first lunar rocks for scientific study through an unmanned, purely robotic mission. However, the United States won that prize, too. *Apollo 11* brought back to Earth the first geological samples ever collected on another Solar System body, 44 pounds of certifiably genuine Moon rocks. Soviet landers reached the Moon in following years and returned a few precious grams of lunar soils from regions unvisited by the *Apollo* astronauts.

For a few days in the summer of 1969, we knew there was intelligent life on another world, even if it was just us.

APRIL 3–7, 1972: With evidence for a planetary system similar to our own found deep within van de Kamp's decades of Barnard's star data, the time was ripe for further theoretical advances. The first international meeting devoted to debating various ideas about the origin of the Solar System was held in Nice, France. (Given the chance, scientists have an understandable preference to hold their meetings in pleasant surroundings.) A number of distinguished astronomers presented their personal, often idiosyncratic ideas, but with one exception the theories had a common trait: They assumed that the planets formed out of the same nebula that produced the Sun. The nebular hypothesis of Kant and Laplace had triumphed two centuries after it was proposed.

The one stubborn nonconformist was Hannes Alfvén, a Swedish physicist whose pathbreaking work on the behavior of plasmas led to the award of a Nobel Prize for Physics in 1970. Working with his colleague Gustaf Arrhenius, who had also moved to the University of California, San Diego (UCSD), Alfvén proposed that the raw material for making planets was gathered by the Sun's magnetic field from interstellar space sometime after the Sun was formed. However, astronomers found it difficult to believe that the Sun could have picked up enough mass from passing interstellar clouds to create an entire Solar System, and they were skeptical of Alfvén's insistence that electromagnetic fields dominated the process of planet formation—gravity is the primary agent, in most scientists' view. However, with a Nobel Prize in his pocket, Alfvén could command attention, and together with Arrhenius he published a monograph a few years later detailing their theory. As a measure of the regard with which this particular work was generally held, some scientists amended the authors' names on the spines of their copies of Alfvén and Arrhenius's monograph to read "All-vain and Erroneous."

The Nice symposium brought forth another analysis of van de Kamp's Barnard's star data, performed by a young postdoctoral fellow, David C. Black, at NASA's Ames Research Center in California. Black had just scored a coup in his doctoral work with his advisor Robert Pepin at the University of Minnesota: He had discovered evidence of an anomalous concentration of isotopes of the gas neon in samples of meteorites, a clue that was clearly important for studying the origin of the solar nebula's dust grains. Rather than continuing in a successful field of study, as most scientists inevitably do, Black had gone to NASA Ames and immediately switched fields from nuclear chemistry to Solar System formation.

Black argued at Nice that the best explanation for van de Kamp's data was a system of two or even three planets, with the best fit also requiring that the planets *could not orbit within the same plane.* If this was true, then Barnard's star once again had a planetary system that did not resemble ours at a fundamental level: Its planets would not orbit in a single plane, the characteristic that points persuasively to our planets having formed out of a flattened disk. In the discussion period after Black presented his results, another scientist pointed out that systems of binary stars often contain stars that orbit in very different planes, implying that perhaps a stellar system had been found, rather than a planetary system. Black disagreed and concluded that we should not be surprised to find planetary systems different from our own, and this view seemed to carry the day.

JUNE 1973: Van de Kamp could not be bothered by Black's new orbital solution—something much more troubling was looming on the horizon. Van de Kamp's colleague at the Sproul Observatory, John L. Hershey, had just published a detailed analysis of the position of a star

(Gliese 793, the 793rd star on a list of nearby stars compiled by the German astronomer Wilhelm Gliese) that had been observed since 1937 with Sproul's 24-inch refracting telescope.

Hershey had reanalyzed the Gliese 793 data with a new automatic plate measuring machine at the U.S. Naval Observatory, developed there by Strand, who had studied 61 Cygni years before when he was still at Sproul. Hershey found something that must have been unsettling to van de Kamp. The 30 years of data for Gliese 793, whether obtained by manual or automatic measurement, showed a wobble that was suspiciously similar to that in Barnard's star. When Barnard's star moved in one direction in 1949, so did Gliese 793. When Barnard's star moved back in the other direction in 1957, so did Gliese 793. The situation was rather like the Harpo Marx and Lucille Ball routine in which Lucy, wearing a Harpo costume, tried to fool Harpo into thinking she was his mirror reflection by precisely duplicating his motions. It did not work for Lucy, and it would not work for Barnard's star and Gliese 793. Either both stars had exactly the same set of planets pulling them back and forth, at exactly the same time and in exactly the same direction, which was clearly preposterous, or else something was very wrong indeed.

Both Barnard's star and Gliese 793 had been observed with the same telescope at Sproul for decades. Hershey calmly noted that the apparent shift in position of Gliese 793 in 1949 coincided with the installation of a new piece of equipment on the 24-inch refractor and with a change in the photographic plates being used. Similarly, the jump in 1957 coincided with another adjustment of the lens of the telescope. Evidently these seemingly innocuous changes had introduced a tiny error that could only be detected after years of observations. Once these erroneous jumps were removed, the evidence for a planetary system around Barnard's star might well evaporate.

The enormous implications of Hershey's revelation for van de Kamp's discovery were not mentioned anywhere in the published paper. In 1972 van de Kamp had stepped down as director of the Sproul Observatory and retired as a professor at Swarthmore College, but he remained at Swarthmore nevertheless. In the name of intellectual honesty, Hershey's analysis had to be revealed to the outside world, but in deference to his senior colleague and former director, Hershey left it to van de Kamp to sift through the debris of the Barnard's star data. This must have been a painful period for the group at Sproul.

But there was much more bad news to come in 1973. A separate team of astronomers had followed the motion of Barnard's star in an attempt to confirm van de Kamp's planets. Any controversial or delicate measurement in science must be replicated by another research team before its validity can be fully ascertained—the chances of two different research teams making the same mistakes and fooling themselves are generally small, particularly if different techniques are used to study the same problem. Scientists are a critical bunch, by training if not by nature, and usually do not mind finding errors in the work of others—some even seem to take delight in this task.

The new study of Barnard's star came from astronomers George Gatewood, of the University of Pittsburgh's Allegheny Observatory, and Heinrich Eichhorn, of the University of South Florida (USF) in Tampa, Florida. Using 241 photographic plates taken at the Allegheny Observatory in Pennsylvania and the Van Vleck Observatory in Connecticut between 1916 and 1971, Gatewood and Eichhorn measured the location of Barnard's star using the same automatic machine at the U.S. Naval Observatory that Hershey had used. They purposely did not use any plates taken with the Sproul refractor. The work was Gatewood's Ph.D. thesis research; Eichhorn served as the Ph.D. advisor, as there was no suitable professor at Pittsburgh.

In the winter of 1972–1973, Gatewood worked in Eichhorn's office in the Physics Building at USF to reduce his data. Gatewood and Eichhorn made for a contrasting pair: Gatewood was tall and thin, with a soothing voice and demeanor, while Eichhorn was tall and wide, with an excitable Austrian accent and a merry swagger. Eichhorn had a European approach to astronomy, preferring not to let astronomy get too much in the way of enjoying life. On the other hand, Gatewood was glad to be working in astronomy again, having taught junior high and worked as a waiter to support himself, and he threw himself into the task. He would work all weekend long at the USF computer center to find the best fit to the data, taking sole control of the campus's fastest computer and sleeping on a table while the computer churned away.

While Gatewood performed his labor of love in the Physics Building, several floors below I happened to be busily taking my senior year classes in mathematics and physics and wondering where to go to graduate school. I was totally immersed in physics at the time and largely oblivious to the workings of the astrometrists—I never took an astronomy course as an undergraduate, and about all I knew was that there was a group up there that worked on determining binary star orbits. Compared to physics, which at the time still held the promise of ever higher energies and ever smaller constituents of matter to discover, the analysis of binary star orbits seemed hopelessly antiquated.

Gatewood and Eichhorn subjected their raw data to a more precise analysis than van de Kamp had used, the best that could be performed given the data—Eichhorn had literally just written the book on precision astrometry and knew what had to be done. They could show that any errors left by their analysis were small enough that van de Kamp's planet should almost leap out of the data.

It did not. Gatewood and Eichhorn found no sign of even a single planet orbiting Barnard's star, much less two or three. Gatewood and Eichhorn published their results in October 1973, right on the heels of Hershey's paper, also in the *Astronomical Journal*. They concluded their paper by pointing out that there had been other instances "when astrometric investigations have suggested the reality of actually unreal things," a phrase that must have been written by Gatewood, who conducts himself as a courtly gentleman. There was no hint of gloating over the demise of van de Kamp's extrasolar planet: in typical, understated, astronomical style, the abstract of the 1973 paper merely noted that Gatewood and Eichhorn's analysis had "failed to confirm van de Kamp's published orbit."

Van de Kamp was not about to concede defeat on Barnard's star, however. He had put in too many decades of effort to give up easily. Van de Kamp was an extraordinarily patient, persistent man, as he had to be—he would salvage what he could from the decades of old data and continue on with gathering new, uncorrupted observations of Barnard's star.

In two more years van de Kamp was ready to publish a new analysis of Barnard's star, using just the photographic plates taken since the 1949 telescope upheaval. The wording of his August 1975 *Astronomical Journal* paper was carefully chosen—the systematic errors caused by the changes to the Sproul telescope in 1949 were variously referred to, in the euphemistic jargon of the astrometrist, as "systematic instrumental equations" or simply as "systematic effects." The 1957 telescope change uncovered by Hershey was judged not to have an effect on Barnard's star, so all the plates in the interval from 1950 to 1974 were used.

Van de Kamp found that he still had evidence for a wobble of Barnard's star that could be explained by the presence of two planets on circular orbits only slightly changed from before—an inner 1 Jupiter-mass planet orbiting at 2.7 times the Earth-Sun distance, and an outer planet with a mass 40 percent that of Jupiter at 4.2 times Earth's distance. Van de Kamp admitted that "systematic error" had caused the reduction of the mass of the outer planet from 1.1 times the mass of

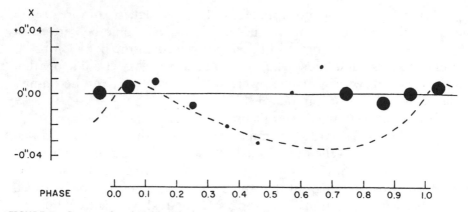

FIGURE 7. Gatewood and Eichhorn's data on Barnard's star, showing the disagreement of their observations of the star's wobble in one direction with that predicted by van de Kamp's planetary solution (dashed line) over one orbital period. Size indicates degree of confidence. (Reprinted, by permission, from G. Gatewood and H. Eichhorn, 1973, *Astronomical Journal*, volume 78, page 775. Copyright 1973 by the American Astronomical Society.)

Jupiter to 0.4. However, throwing away the data from before 1950 had little qualitative effect on his 1975 result—van de Kamp still believed he had evidence of a system of two Jupiter-like planets on circular orbits. The year before, van de Kamp had published a claim of finding evidence for a 6 Jupiter-mass object in a 25-year orbit around the similarly nearby star Epsilon Eridani, which was duly noted as a new "unseen companion" several months later in *Sky and Telescope*. Van de Kamp's faith in his work remained strong despite the misgivings of others.

But the game was over—van de Kamp's arguments were not generally accepted by his colleagues. The disastrous revelations of 1973 caused van de Kamp to lose much of his credibility on Barnard's star, though he had made many other important and lasting contributions to the astronomy of low-mass companions and binary stars, received numerous awards and medals, and even served as the first Director of the Astronomy Program at the National Science Foundation (NSF). By 1976, the astronomy textbooks were being rewritten to state that there was no firm evidence for any extrasolar planets after all. The field of extrasolar planet searches fell into a deep sleep.

4

MAKING PLANETS
FOR FUN AND PROFIT

Give me matter and I will construct a world out of it.

—Immanuel Kant (1724–1804)

The search moved back to square one. Extrasolar planet research once again became the sole dominion and playground of the theorists, unencumbered by any constraints or clues about what the planet formation process should produce save our own Solar System. With only one known solar system, the theorists' game began to explain, in as much mind-boggling and fanciful detail as possible, how our Solar System could have formed, and by analogy perhaps how other planetary systems formed as well. The underlying assumption was that whatever happened to make our Solar System almost certainly happened elsewhere in the universe, and hence that planetary systems should be reasonably commonplace—any argument that implied that we might be unique was to be considered suspect. The Kant-Laplace hypothesis of coupled star and planet formation surely did not apply only to the Solar System.

The Marquis de Laplace's ideas about how planets might form out of the nebula were necessarily somewhat vague, given the state of science in the eighteenth century. Laplace suggested that as the nebula slowly contracted, rings of gas would be shed at the nebula's equator and left behind in orbit. These rings would then break up into spherical blobs of gas that could contract and heat up and form the planets. While this process might reasonably explain the formation of gas giant planets like Jupiter and Saturn, the formation of the icy outer planets

(Uranus, Neptune, and Pluto) and the rocky terrestrial planets (Mercury, Venus, Earth, and Mars) did not appear to be at all likely in Laplace's scenario.

While Chamberlin and Moulton's tidal theory was discarded soon after Russell pointed out that hot filaments pulled from the Sun would expand and not contract, they did introduce a concept and a word that both survive to the present day: the idea that the planets formed not from the contraction of hot spheres of gas, but from the collisions of cold rocks (*planetesimals*) in orbit about the early Sun. Chamberlin and Moulton called planetary building blocks planetesimals, a word formed by combining the words *planet* and *infinitesimal*. Chamberlin and Moulton imagined that a close encounter between two stars would produce a spray of hot gas that would quickly cool and form small chunks of rock and ice, the planetesimals, whereas Russell had argued persuasively that superhot solar plasma would dissipate and not form any planetesimals at all. The idea of making planets out of planetesimals survived, however, because of its intrinsic reasonableness—building planets up, one collision at a time, seems like a natural way to build a rocky planet, as a single glance at the crater-pocked surface of the Moon confirms.

But where then to make the planetesimals, if not in a thin filament pulled from a star? The answer that made the most sense turned out to be inside Kant and Laplace's nebula, of all places.

Astronomers had learned that the clouds of gas in which new stars were observed to be forming also contain a population of minute solid particles called dust grains. These grains typically are so small (less than 0.1 micron, where 1 micron = one millionth of a meter) that it would take several hundred thousand dust grains to make a strand just an inch long.

While there is no way that astronomers can see such incredibly small objects at distances of hundreds of light years or more, the presence of the dust grains can be deduced from the effects they cause, such as the reddening of starlight. The human eye can see light ranging in wavelength from 0.38 microns (violet) to 0.75 microns (red), where the wavelength is the distance from one crest of a light wave to another, using the analogy of a wave on the ocean. Interstellar dust grains are generally smaller than the wavelength of visible light, and because of this they preferentially scatter (or deflect) starlight of shorter wavelength—longer-wavelength light waves can pass right around them. This means that the blue light is knocked off course and sent in another direction more than red light is. The effect was discovered experimentally by the English physicist Lord Rayleigh in 1871.

Lord Rayleigh proved that scattering of the Sun's light by nitrogen and oxgyen molecules in the Earth's atmosphere is what causes the sky to be blue; the blue color is the blue light that happens to get scattered in the direction of your eye by molecules floating in the atmosphere all over the sky. However, when you look at the Sun as its light passes through more and more of the Earth's atmosphere during a sunset, Rayleigh scattering means that the Sun's blue light is scattered *away* from the direction toward your eye, and the loss of this blue light makes the Sun look much redder than when it is overhead. The same thing happens when a star lies on the other side of an interstellar cloud of dust grains. By the time a star's light reaches the Earth, much of the blue light has been scattered off into other directions, while the red light continues on to Earth, making the star appear to be redder than it would be if it did not happen to lie behind a dust cloud. The amount of reddening of starlight can then be used to infer how much dust, and of what size, lies in the intervening clouds. This is done by calculating what a hypothetical cloud of dust particles would do to the light from a background star with a known brightness, and adjusting the properties of the hypothetical dust cloud until the observed reddening is obtained.

Astronomers could then show that interstellar clouds consist primarily of hydrogen and helium gas, just like the Sun, but with a small percentage of their mass residing in cold dust grains. This recipe defines the cosmic composition. Because interstellar clouds are the starting material for new stars, the nebula that contracted to form the Sun must have been laced with dust grains that could be used to form planetesimals and eventually planets as well.

Theorists thus began to construct models of planet formation based on Chamberlin and Moulton's fruitful concept of planetesimals in orbit around the early Sun. The Irish astrophysicist K. E. Edgeworth was one of the first to do so, publishing his ideas about the origin of the Solar System in the British journal *Monthly Notices of the Royal Astronomical Society* in 1949. Edgeworth suggested that the early Solar System contained a thin disk of planetesimals resembling the rings of Saturn, and that this disk would be able to form clusters of planetesimals through the gravitational forces that act between any two bodies with mass. That is, Edgeworth suggested that the disk would spontaneously break up into a number of aggregations of planetesimals, each of which would then contract to form an even larger body. This process could greatly speed up the process of turning minute dust grains into planets the size of Earth or larger. The process Edgeworth described is termed a *gravitational instability*.

Another type of gravitational instability was advanced by Gerard Kuiper in a paper published two years later in the United States (in the *Proceedings of the National Academy of Sciences*). Like van de Kamp, Kuiper was a Dutch-born astronomer who emigrated to the United States early in life. He noted that the planets, especially the terrestrial planets, contain only a small portion of the hydrogen and helium gas that must have accompanied the dust grains of which they are chiefly composed. He argued that one could estimate the mass of the solar nebula that formed the planets by calculating how much hydrogen and helium gas would need to be added in order to make a nebula of cosmic composition out of the existing planets. Kuiper came up with a total nebula mass in the range of 1 to 10 percent of the Sun's mass; because substantial nebular gas might have been lost during the process of forming the Solar System, these estimates gave only a lower bound on how massive the solar nebula must have been—it could have been more massive yet.

Because these estimates were only lower bounds, Kuiper argued that the solar nebula must have had a mass close to 10 percent that of the Sun. In that case, Kuiper showed that the solar nebula gas may have been massive enough to undergo a gravitational instability of the sort that Edgeworth had suggested occurred in a thin disk of planetesimals—gravity can cause a massive disk of cold gas to clump together in much the same way as a thin disk of solid planetesimals. Kuiper believed that the gravitational instability would form large clouds of gas and dust, which he called protoplanets; the protoplanets would be long lived and capable of forming the known planets. Kuiper further proposed that the unwanted gases that would have accompanied the formation of the terrestrial protoplanets were driven off by heating from the Sun, leaving behind a set of rocky cores. Kuiper's arguments implied that all the planets of the Solar System could be formed through a single mechanism: gravitational instability of a massive, gaseous solar nebula.

Kuiper's 1951 vision was ambitious but ultimately unconvincing to others. Because of one man, Otto Schmidt, a movement was slowly building in the Soviet Union to picture the formation of all the planets as occurring through a competing mechanism—collisions between planetesimals.

The Russian mathematician Otto Schmidt came of age during the Bolshevik revolution in 1917 and rose rapidly in the Soviet scientific ranks. Schmidt's political and scientific successes, including heading the first Soviet air expedition to the Arctic in 1937, led to his becoming the director of a new institute of theoretical geophysics in 1937 and vice-president of the powerful Soviet Academy of Sciences in 1939. His formidable administrative duties might have kept him from doing any further science, were it not for his German ancestry: In 1942, during the horrors of the German invasion of the Soviet Union, Joseph Stalin ungratefully kicked Schmidt out of the Academy job. With considerable free time suddenly on his hands, Schmidt turned again to science, and along with his pupils he began the serious development of the theory of planetesimal accumulation, work that would continue for the next several decades in Moscow.

This pathfinding research was summarized in 1969 in a monumental treatise written by Schmidt's leading pupil, Victor Safronov, entitled *Evolution of the Protoplanetary Cloud and Formation of the Earth and the Planets*. The book promised to deliver a comprehensive framework for the problem of forming planets, and it did. Once it was translated into English and published in 1972, the book quickly became required reading in the West for anyone who wished to study planet formation. Reading Safronov's book in the winter of 1974–1975 as a second-year graduate student in physics directly stimulated my personal interest in star and planet formation and diverted me away from mainstream physics. I was able to absorb in a short while the analytical insights that had taken Schmidt and Safronov and their colleagues decades of effort to discover. I could also see that the problem of stellar and planetary system formation was ripe for solution by the increasingly powerful methods of computational physics. The advent of the modern age of planetary formation theory can be traced largely to the arrival of Safronov's book.

Safronov outlined the phases that must occur in order for a planetary system to grow out of a disk of interstellar dust grains. Safronov simplified the problem considerably by completely ignoring the formation of the Sun. He simply assumed that, somehow, the Sun managed to form out of the primordial cloud, and soon thereafter found itself surrounded by the vestiges of its formation process—the protoplanetary cloud. The starting conditions thus were assumed to be a young star with the mass of the present-day Sun, encircled by a flat disk of gas and dust moving nearly on the same Keplerian orbits that characterize the motions of the present-day planets around the Sun. The amount of gas and dust in the disk and its distribution with radius from the Sun could be estimated by working backward from the present-day Solar System, as Kuiper had done. The starting point for planetary formation could thus be specified in some detail, and with this assumption tremendous

FIGURE 8. Victor S. Safronov, the Soviet pioneer of the theory of planet formation. (Courtesy of V. S. Safronov.)

progress could be made as scientists used the known laws of physics and chemistry to predict what would happen next.

Safronov estimated that the mass of the protoplanetary cloud was about 5 percent of the Sun's mass, 1 percent of which he assumed would be in the form of interstellar dust grains. The first phase of evolution had to involve the growth of dust grains to sizes much larger than their initially microsopic sizes.

The process through which this happened is straightforward—collisions between the dust grains led to their sticking together, provided that the dust grains collide in the first place. However, dust grains are few and far between, even in the relatively dense protoplanetary cloud, so growth by this process alone is slow. Safronov realized that because of the downward pull of a portion of the Sun's gravity (toward the disk's middle), the dust grains would settle out of the gas and form a thin disk at the center of the protoplanetary cloud, just as suspended particles in a pond slowly sediment down and form a layer at the bottom of the pond. The "bottom of the pond" in the protoplanetary disk is actually the middle of the disk, because particles above the middle were pulled downward, while those below the middle were pulled upward, meeting in the middle to form a thin dust disk inside the gas disk.

During this settling process, the larger grains moved downward faster than the smaller grains and swept up any smaller grains they

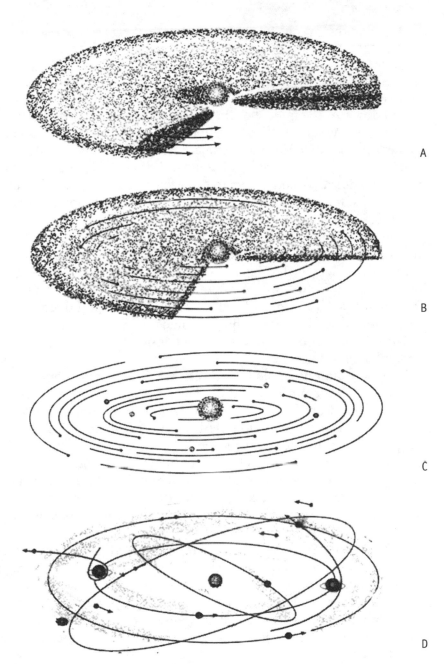

FIGURE 9. The formation of the terrestrial planets involved four distinct phases: (A) Dust grains in the solar nebula coagulated into milewide planetesimals. (B) The planetesimals collided to form Moon-sized planetary embryos. (C) The remaining solar nebula gas and dust was removed from the system. (D) The planetary embryos smashed together in tremendous impacts and formed the final planets, moving on nearly circular orbits. (Reprinted, by permission, from *An Integrated Strategy for the Planetary Sciences: 1995–2010*. Copyright 1994 by the National Academy of Sciences. Courtesy of the National Academy Press, Washington, D.C.)

encountered, adding to their mass. The dust grain coagulation process was also enhanced by the increasing density of grains as they all sedimented down to the middle of the cloud, simply because collisions are more frequent wherever there are more dust grains, as any urban automobile driver knows. Safronov estimated that within about 1,000 years the dust grains at the Earth's position would have sedimented down to form a thin dust disk and grown to the size of children's marbles. By astronomical standards, where stars like the Sun take billions of years to grow old, a thousand years is a blink of the eye.

The second phase occurred even faster, according to Safronov, because the dust disk would be subject to the gravitational instability first proposed by Edgeworth. This instability occurs rapidly, requiring only the time needed for the disk to rotate a few turns. In less than a decade, then, the marbles orbiting at Earth's distance would pull themselves together into loose clumps weighing as much as a several-mile-wide asteroid (that is, about a hundred trillion pounds, the same as about thirty billion automobiles).

The loose clumps would have a very low density when first formed, much like a fragile puffball of dust. Their density would increase with each collision of one clump with another, like automobiles being compacted in a cosmic Demolition Derby, because of the velocity of the clump-on-clump impacts. By the time that each clump had collided with another three or four clumps, the combined body would have been about as dense as a typical rock or chunk of ice, and the clumps would have been 10 miles across. These comet-sized objects were obvious candidates to play the role of the planetesimals postulated by Chamberlin and Moulton.

Planetesimals 10 miles across were big enough that their gravitational attraction for each other drove their subsequent growth—no longer did growth depend solely on two bodies hitting each other and sticking together. If by chance two planetesimals passed close together but did not hit, the gravitational forces between them would deflect their orbits and increase their chances of hitting some other planetesimal sooner or later.

The third phase of growth thus involved collisions between planetesimals large enough that their gravitational forces needed to be included in the analysis. Safronov used a theory developed to study the coagulation of raindrops in the Earth's atmosphere to predict how the planetesimals would grow. The great advantage of coagulation theory was that it was simple enough that Safronov could solve the mathematical equations by hand, without having to use a computer; computers were not readily available for this sort of research at Schmidt's institute in the 1950s and 1960s. Safronov found that a swarm of planetesimals orbiting at the Earth's location would coagulate into a planet with the mass of the Earth in about 100 million years. Using the same coagulation model, Safronov also calculated that even if the hypothesized gravitational instability of the dust disk did not occur, marble-sized solids could grow to the 10-mile-sized planetesimals required for the third phase of growth in about 10,000 years. Both of these time scales seemed perfectly acceptable, given that the age of the Solar System is about 4.6 billion years—Earth still formed rather quickly in comparison.

Because of their gaseous composition, Safronov pointed out that the formation of the giant planets required a fourth phase: Enough gas from the protoplanetary cloud had to fall onto the solid bodies formed during the third phase to account for their hydrogen- and helium-rich outer layers. Safronov estimated that the nebular gas would begin flowing onto the protogiant planets once they had grown to one or two Earth masses in size. Prior to a protoplanet growing to this critical size, nebular gases would form only a relatively thin atmosphere on the protoplanet, but after a protoplanet reached the critical mass, this atmosphere would suddenly collapse downward. More gas would flow in from the nebula to replace the lost upper atmospheric gas, increasing the mass of the protoplanet, and in turn making even more gas flow in, and so on. The whole process would turn into a runaway implosion until the nearby nebula ran out of gas. The time required for the growth of Jupiter was less clear than it was for Earth, but even if it was a bit longer than Earth's 100 million years, that did not appear to present any particular problem.

That was not the case for Uranus and Neptune, however. Safronov found that it would take 100 billion years for them to grow to their present size using his coagulation theory. Coagulation slows down considerably far away from the Sun because of the small number of planetesimals out there and because of their long orbital periods (Neptune takes 165 times as long as the Earth to complete one trip around the Sun). Needless to say, this result was an embarassment to the theory, since 100 billion years greatly exceeds not only the age of the Solar System, but the age of the universe as well. According to coagulation theory, Uranus and Neptune did not yet exist, which would have been a surprise to Herschel and Galle. Safronov suggested solving this critical problem by invoking the presence of 10 times as many planetesimals in the outer Solar System as were needed to make the outer planets, so that Uranus and Neptune could grow much faster. The ad hoc nature of this solution was not satisfying, however. Nevertheless, Safronov's theory seemed to do a reasonably good job of accounting for the formation of the terrestrial and giant planets, even if there were still a few kinks remaining to be ironed out.

Soon after 1972, George Wetherill of UCLA read Safronov's book and was captivated by the problem of planet formation by collisions between planetesimals. Wetherill had a distinguished background as one of the pioneers of using radioactive nuclei to date the ages of rocks, but he dropped that research in order to concentrate on making planets. He was aided by his recent experience in studying the orbital dynamics of asteroids.

Wetherill knew how to use UCLA's IBM mainframe computer to calculate the orbital evolution of bodies moving around the Solar System, knowledge that was critical to making the next major leap in understanding—Safronov and his group were largely left behind at this point because of their continued concentration on problems that could be solved without computers. With the help of computers, many new and important physical effects could be studied, such as allowing the planetesimals to interact with each other as if they were really in orbit about the Sun, rather than as if they were raindrops falling to Earth. Permitting the planetesimals to interact would require a treatment of the orbital motion of each planetesimal, so Wetherill developed a computer code that was capable of handling this and many other complications. Wetherill's approach was to solve the equations governing orbital motion in an approximate fashion. The approximation was called the Monte Carlo technique because it used chance to determine the outcome at crucial junctures, like a gambler trying to decide when to quit by monitoring his or her luck at the roulette wheel.

Wetherill adopted the Roman method of divide and conquer in order to make progress. He recognized that the phase of growth that Safronov treated with coagulation theory was really two phases: an early phase, in which coagulation theory was still a reasonably good choice, and a later phase, in which the orbital motion of the planetesimals had to be included. In the early phase, the planetesimals in the inner Solar System were so numerous (numbering in the trillions) and hence close together that the fact that they were orbiting the Sun made little difference—they might as well be raindrops running into each other. In the later phase, the planetesimals had grown into a much smaller number of much larger bodies, up to the size of the Earth's Moon, moving on roughly circular orbits, and these relatively isolated bodies would only collide if their orbits were disturbed by their continual pulling and yanking on each other, like Neptune on Uranus's orbit. The problem of the final assembly of the Earth was ripe for attack by Wetherill's Monte Carlo computer codes.

Like Safronov skipping over the question of the Sun's formation, Wetherill skipped over the question of the early phase of planetesimal collisions and tackled the problem of the final phases of planetary formation in the inner Solar System. To reproduce the terrestrial planets, Wetherill started his computer calculations with a collection of several hundred bodies moving on circular orbits about the Sun around the orbits of Venus and Earth. The total mass of the initial bodies was taken to be about that of the present terrestrial planets, so that each body started with about a lunar mass (the Moon weighs about 81 times less than the Earth). Wetherill was then able to show that this system of bodies would slowly perturb each other's orbits, over tens of millions of years, turning their initially circular, nonintersecting orbits into oval (elliptical) orbits, in which the bodies would eventually collide and grow larger.

Wetherill's Monte Carlo calculations showed that there was a good chance for this collection of massive planetesimals to evolve into our present Solar System. Because of the random, unpredictable nature of planetesimal collisions, Wetherill had to run his models over and over again, slowly building up enough statistical knowledge to predict the outcome in general. He often found that two Earth-mass planets were formed (Venus and Earth), as well as one or two smaller planets (Mercury and Mars), orbiting more or less at the locations of the terrestrial planets in our Solar System. The whole process took about 100 million years to complete, as Safronov had predicted.

In 1975 Wetherill left UCLA to become director of the Carnegie Institution of Washington's Department of Terrestrial Magnetism (DTM), in Washington, D.C. Wetherill had been a staff member at DTM before being lured westward to UCLA in 1960, at a time when the rapidly growing UC campuses were raiding other universities in search of professors. Wetherill was glad to escape his brilliant but often domineering DTM director, Merle Tuve. When he returned as DTM's director, Wetherill made it a point not to meddle in his staff's research.

In spite of his administrative responsibilities as DTM's director, Wetherill continued to carry forward his research program on planet formation. He also brought in a few new people to work on the problem, notably Stuart Weidenschilling in 1976, fresh from completing his Ph.D. research on planet formation at the Massachusetts Institute of Technology.

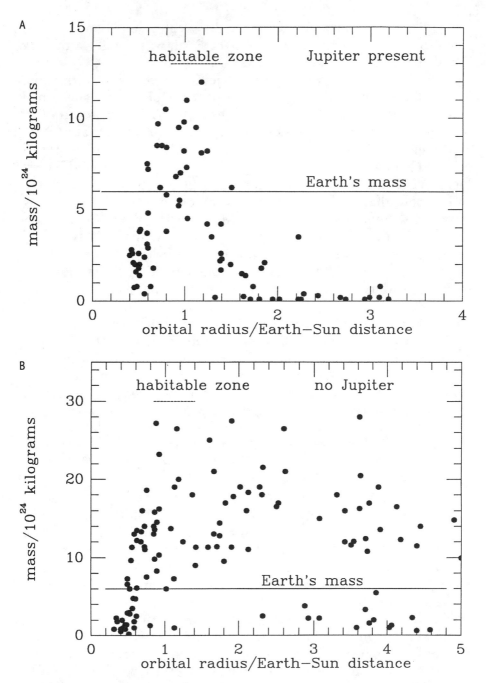

FIGURE 10. Wetherill's calculations of the formation of the terrestrial planets through the impacts of lunar-sized planetary embryos. Because of the random nature of the collision process, the results can only be discussed in a statistical sense. Here, the results of 23 simulations (A) and 34 simulations (B) are presented. (A) shows what happens in the presence of Jupiter, and (B) in the absence of Jupiter. (Adapted from G. W. Wetherill, 1996, *Icarus*, volume 119, pages 226 and 235.)

During his two-year postdoctoral fellowship at DTM, Weidenschilling worked on the first two phases of planet formation, the phases of growth from dust grains to miles-wide planetesimals. He studied the interactions of small planetesimals with nebular gas, interactions that could have a profound implication for Safronov's hypothetical dust disk instability (possibly preventing it).

The idea of dust disk instability had arisen not only in Ireland and in the Soviet Union, but also in southern California. Prior to the publication of the English edition of Safronov's book, the instability had been rediscovered and studied in detail by Peter Goldreich of the California Institute of Technology and his Ph.D. student, William Ward. Goldreich and Ward decided in 1972 that the gravitational instability of the dust disk would produce mile-sized planetesimals in a few thousand years, in good agreement with Safronov's analysis of the final outcome. Dust disk instability became known in the West as the Goldreich-Ward instability, in spite of the earlier work.

Weidenschilling pointed out in 1977 that the gas in the nebula would rotate around the Sun slightly slower than the solid bodies. This happens because the gas is partially supported against falling inward onto the Sun by collisions with gas molecules closer to the Sun. This means that the gas effectively does not feel as strong a gravitational pull inward as the solid bodies do. As a consequence, the gas must rotate slower, so that a smaller centrifugal force outward is needed to balance the reduced inward pull of the Sun. The resulting slight difference in velocity between the gas and the solids might be large enough to keep the solids suspended in the gas, just like stirring a cup of tea with loose tea leaves settled at the bottom lifts the tea leaves and swirls them about the cup. Weidenschilling suspected that the dust disk could never become dense enough to undergo the gravitational instability. Such a high density would require the solar nebula to contract to a thinness of only about 100 miles thick, at a distance 93 million miles from the Sun, a daunting prospect. Like Safronov, Weidenschilling pointed out that the alternative of allowing solids to grow through simple collisions to cometary-sized planetesimals might work just as well as the Goldreich and Ward instability.

While this attention was focused on the formation of Earth and Venus, the problem of Jupiter formation was also under scrutiny. Kuiper's idea of making giant planets through gravitational instability had been superseded by Safronov's vision, but it soon became obvious that a major problem had developed with Safronov's approach to giant planet formation—it was likely to run out of gas.

Because the process of building planets through collisions occurs slower and slower farther away from the Sun, it seemed that Jupiter could not form by this process any faster than the Earth did, which took about 100 million years. However, by the late 1970s astronomers had been able to estimate the ages of young stars and had found that none were much older than about 10 million years. After 10 million years, the stars became normal, middle-aged stars like the Sun—their protoplanetary disks were expected to be gone, and there would be little or no disk gas left to build the gas giant planets. There was no problem in taking 100 million years to assemble the Earth, because the Earth showed no signs of ever having had a massive atmosphere of hydrogen and helium; but Jupiter and its slowpoke siblings presented a real problem.

There was another good reason that Jupiter had to be made fast: so it could serve as a disruptive influence on the growth of planetesimals in the region between it and Mars. If Jupiter did not form early, there was a danger that the accumulation models would predict the existence of a planet in the Solar System between Mars and Jupiter where one does not exist today, an obvious failure to be avoided. The region between Mars and Jupiter is the home of the asteroid belt, a motley collection of thousands of bodies derived from planetesimals that failed to smack into a protoplanet. The asteroids are in the process of grinding themselves down to dust grains again through high-velocity collisions, and taken together the asteroids do not have enough mass to make a decent planet. Something must have interrupted the planet formation process in the asteroid belt, and Jupiter is generally considered guilty until proven otherwise.

A way to make Jupiter quickly had been suggested by Canadian-American astrophysicist Alastair G. W. Cameron, one of the most prolific and protean thinkers about Solar System formation during the 1960s and 1970s. Cameron had suggested in early 1972 (independent of Safronov) that the giant planets had formed by pulling gas onto a solid core formed through collisions of planetesimals. He estimated that the critical mass needed to suck in substantial nebular gas was about 10 Earth masses, which would take even longer to form than Safronov's one or two Earth masses. However, Cameron changed his mind about giant planet formation a few years later, when he began to compute models of how the gaseous portion of the disk would evolve. The disk models relied on a concept introduced by two British astrophysicists.

A theoretical model (i.e., a set of mathematical equations chosen to mimic real physical processes) was published in 1974 by Cambridge University astrophysicists Donald Lynden-Bell and James Pringle. The model predicted how disks made up of viscous fluids, like corn syrup, would change with time. Lynden-Bell and Pringle's concept of a disk with viscosity neatly solved the Laplacian problem of the Sun's excess angular momentum. A viscous disk moves angular momentum outward, away from the Sun, while at the same time moving most of the disk's mass inward, onto the Sun. Like Goldreich and Ward, Lynden-Bell and Pringle generally received credit for the concept of the viscous disk, though its antecedents date back to the work of von Weizsäcker and that of his German colleague, R. Lüst, in the 1950s. (Being second to publish scientific results, but in the English language, at times can turn out to be more important than publishing first in a non-English language, especially if the later paper also presents an improved analysis of the problem that may help justify a claim for precedence.)

Al Cameron, known in much of the astronomical community simply as Big Al, prided himself on his ability to accept fresh new ideas and to relinquish those that had grown stale. Cameron was the first to apply Lynden-Bell and Pringle's viscous disk theory to the problem of Solar System formation. The models he calculated in 1977 implied that the effects of viscosity might lead to disks so massive and so cool that they would be subject to gravitational instability. Cameron predicted that the disk would first break up into rings, and then the rings would collapse to form giant gaseous protoplanets. The outcome was the same as that advanced by Kuiper in 1951, though Cameron apparently was unaware of Kuiper's earlier work.

Cameron and William DeCampli, Cameron's graduate student at the Harvard-Smithsonian Center for Astrophysics, thereafter began to study the properties of giant gaseous protoplanets. In 1979 they calculated that the dust grains inside the protoplanet could rain down to the center of the protoplanet and form a core of iron, rock, and ice. This was a key finding, because NASA's robotic missions to the outer planets had helped to show that Jupiter and Saturn have rock and ice cores about the same size as Uranus and Neptune (that is, about 10 Earth masses). The process of dust grain rainout was crucial for making giant planets by the gravitational instability mechanism, for without it the protoplanets most likely would not have dense cores. In spite of this promising start, a few years after completing his Ph.D. DeCampli left astrophysics to attend medical school and became a cardiologist.

The remarkable fact that Jupiter, Saturn, Uranus, and Neptune each seemed to be built around a solid core with a mass of about 10 Earth masses was not easily explained in the context of the giant gaseous protoplanet theory—there was no particular reason why these four planets should have similar core masses if they formed by gravitational instability.

An extremely attractive explanation for the similarity in core masses of the giant and outer planets arose from the work of the Japanese school of astrophysicists organized and led by Chushiro Hayashi at Kyoto University. Hayashi had been a pioneer in the early theoretical work on young stars, and in his later years he turned his attention to the question of planet formation. Hayashi dominated his group—I remember seeing Hayashi at a conference in 1979 and watching his group members walk quietly a few steps behind their leader and sit one row behind him in the auditorium. Hayashi believed that the Earth, and not just Jupiter, had formed while the solar nebula was still around, so his students were predisposed to worry about the effects of nebular gas.

One of Hayashi's Ph.D. students, Hiroshi Mizuno, discovered in 1980 that the similarity in core masses made perfect sense if one forgot about gravitational instability once again and went back to the alternative of building a core and then accreting disk gas onto the core. Mizuno's detailed models of the interaction of a solid core with the disk gas showed that once the solid core exceeded about 10 Earth masses, the disk gas would begin to flow rapidly onto the core and would drown it in hydrogen and helium. While Cameron had come up with about this same value for the critical core mass beforehand, Mizuno had calculated it in some detail, and, more remarkably, found that the critical core mass did not vary with distance from the Sun: Jupiter, Saturn, Uranus, and Nepture should therefore all have about the same core mass according to Mizuno's calculation. They did indeed, and so Mizuno's calculation was taken as a strong argument in favor of making giant planets by the core accretion method. The nagging problem of making the core fast enough remained, however.

The pendulum was given another push back in the direction of the core accretion method through the work of Douglas Lin and John Papaloizou. Lin, of UC Santa Cruz (UCSC), and Papaloizou, then of Britain's Institute of Astronomy in Cambridge, had complementary traits that made for a finely balanced team. Lin was a rapid-fire speaker known for flipping quickly through overhead transparencies with unlabeled figures during his talks. He was frequently in attendance at conferences, where he could push his latest ideas. Conversely, Papaloizou was a quiet astrophysicist with strong mathematical skills. Papaloizou generally avoided meetings, and when forced to attend, he would sit in the back of the room carrying on with his hand-written calculations.

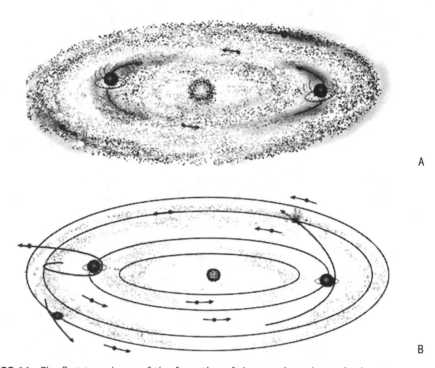

A

B

FIGURE 11. The first two phases of the formation of the gas giant planets in the outer Solar System were the same as in the terrestrial planet region, except that the planetary embryos grew to 10 Earth-mass size before the solar nebula disappeared. The last two phases were (A) rapid accretion of solar nebula gas by the planetary embryos, and (B) Removal of the remaining gas after the giant planets had reached their final masses. Giant planets are believed to form on circular orbits. (Reprinted, by permission, from *An Integrated Strategy for the Planetary Sciences: 1995–2010*. Copyright 1994 by the National Academy of Sciences. Courtesy of the National Academy Press, Washington, D.C.)

Lin and Papaloizou published a viscous disk model in 1980 that yielded a result quite different from Cameron's regarding the gravitational instability mechanism. Lin and Papaloizou found no tendency for their viscous disks to be massive enough to form giant protoplanets spontaneously. They also stated that the maximum mass of a giant planet is about the mass of Jupiter. This conclusion was based on their analysis that once a giant planet formed by the core accretion mechanism had grown to Jupiter's mass, its gravity would begin to clear a gas-free gap in the disk around the planet, thereby shutting off the flow of any additional gas onto the planet. Our Solar System thus had managed to form about the biggest planet that it possibly could have; there was little hope of finding extrasolar planets with substantially larger masses, planets that would be considerably easier to detect.

A more ominous prediction about giant planets was made about the same time by Peter Goldreich and his former Caltech postdoctoral fellow, Scott Tremaine, by then at the Institute for Advanced Study in Princeton, New Jersey. Goldreich was a theoretician of legendary brilliance, one of those rare scientists who can and do make fundamental contributions in many different areas of their science. Tremaine rapidly was becoming known as one of the world's leading astrophysical theorists. Goldreich and Tremaine constituted a potent theoretical attack squad, a sort of cosmic SWAT team, ready to handle any conundrum the universe could throw at them.

Motivated by a desire to explain the peculiar structure of the rings around Saturn, Goldreich and Tremaine had been studying the gravitational interactions of Saturn's inner satellites with the ring particles, both of which orbit around Saturn in the same direction and in the same orbital plane. In many ways, the Saturn system is a mini-Solar System, with Saturn playing the role of the Sun, the satellites pretending to be planets, and the ring particles standing in for the gas and dust of the solar nebula. Goldreich and Tremaine found that Saturn's satellites could force the ring particles to follow orbits such that the rings resembled the spiral arms of tightly wound spiral galaxies. However, the spiral arms would also have an effect on the satellites and could make the satellites themselves spiral either inward or outward to new orbits. Goldreich and Tremaine then applied their equations to the case of Jupiter and the solar nebula and predicted that the resulting interactions would cause Jupiter either to spiral inward onto the Sun or outward to oblivion in about 10,000 years. Jupiter's orbit would be forced to become circular by the same forces, in the even shorter time period of a few hundred years. The disquieting idea arose that after the Jupiters were made, they just might disappear into their suns and be lost forever. Theorists kept this frightening prospect quiet—obviously, our Jupiter had somehow survived, but for all we knew it was a fluke, the one exception.

The swinging pendulum of ideas about Jupiter formation had reached the end of another arc, and giant planet formation by gravitational instability was unfashionable once more. The perennial problem of figuring out how to make Jupiter fast enough was publicly bemoaned, but not solved. Even though it was again the leading idea, the popularity of Mizuno's mechanism for giant planet formation was based implicitly on the faith that *somehow* a solution to the time scale problem would be found.

After the meltdown of van de Kamp's claims for Barnard's star in 1973, theorists had a rare opportunity to figure out how other planetary systems form and what they might look like, before any genuine extrasolar planetary systems were found. They had the chance to make striking theoretical predictions that might later be verified, like the celebrated predictions of Adams and Le Verrier and Galle's subsequent discovery of Neptune. They also had the chance to be proven wrong if their predictions came to naught. But because of the focus on our own Solar System, little serious effort went into making predictions about other solar systems. Theorists by and large played it safe and were content to try to solve a problem for which much of the answer had been known for centuries.

5

I Thought I Saw
a Brown Dwarf Last Night

*In the field of observation, chance only favors those minds
which have been prepared.*

—Louis Pasteur (1822–1895)

January 1978: Kant and Laplace had hypothesized centuries before
that the entire Solar System was created out of a single primordial
cloud, and their arguments were persuasive to nearly all who worked
on planet formation in the 1970s. Only Alfvén, whose willingness to
ignore conventional wisdom led to his receiving a Nobel Prize, had
proven contrary, proposing that the planets formed out of dust gathered by the mature Sun's magnetic field. The nearly universal agreement
that planet formation was inextricably linked to star formation was
formally recognized by a scientific conference held in Tucson—the first
Protostars & Planets conference.

The motivation for this interdisciplinary conference was the realization that the problems of star formation and planet formation had to
be solved in unison. It would not do to have a group of astronomers
and astrophysicists talking only to themselves about how stars are
formed, while somewhere else planetary and meteorite scientists were
trying to decipher the puzzle of Solar System formation. Surely both
groups had important things to learn from each other, and the best way
for this to occur was to hold joint meetings.

One of the first things conference participants learned was the difference in their cultures, as people marched up to the microphones in
the center of the aisles to ask questions of each speaker. Astronomers,

who tend to belittle their rivals' ideas more through understated innuendo than through outright accusations of wrongdoing, watched the more combative meteoriticists figuratively go at each other's throats, cigars still clenched in their teeth.

Nevertheless, the Protostars & Planets conference marked the emergence of a community of scientists who would approach the problems of star and planet formation from a common direction. Even Alfvén had come around, dropping his non-Laplacian ideas and presenting a model that envisioned planets and stars forming together out of the same region of dusty plasma. Alfvén still insisted, however, on a much larger role for electromagnetic forces than others considered to be likely, and he spent most of the conference sitting by himself in the front row. The conference produced no new converts for his cause.

JULY 1983: The Protostars & Planets conference delicately sidestepped the question of detecting extrasolar planets and other low-mass companions to stars—the memory of the Barnard's star disaster was still too raw. It was not until five more years had passed that the next step was taken, by Robert Harrington, an astronomer at the U.S. Naval Observatory in Washington.

He did not fully know it at first, but in mid-1983 Harrington found evidence for something almost as elusive as an extrasolar planet. Harrington had found the astrometric signature of a *brown dwarf,* a hypothetical object that hovers in the netherland between the realm of stars and the realm of planets.

Harrington was looking for hidden companions to stars using data taken with the U.S. Naval Observatory's 61-inch-diameter reflecting telescope near Flagstaff, Arizona. The observations began in earnest in 1972 when he had taken charge of the Naval Observatory's program to measure the distances to nearby stars. One of the first things Harrington did that year was to add Barnard's star to the list of stars to be scrutinized. At that time Barnard's star had not yet been demoted to a planetless star, and besides, Harrington had been a student of van de Kamp's at Swarthmore when the first announcement about Barnard's star was made in 1963—an event that electrified the campus. In a small college like Swarthmore, an eminent astronomer like Peter van de Kamp stood out, and after van de Kamp announced his detection, he was viewed with awe by the students. Though he earned a degree in physics from Swarthmore, Harrington had helped out with van de Kamp's observations, made each clear night on campus at the 24-inch telescope. Harrington was well suited to follow in van de Kamp's footsteps.

Harrington's program sought to measure the distances to nearby stars by observing the semiannual shift of the stars' apparent locations in the sky. Taking photographic plates six months apart in time means they are taken on opposite sides of the Earth's 12-month-long orbit around the Sun, at locations on the ends of a line passing through the Sun and with a length equal to the diameter of Earth's orbit, about 186 million miles. This distance is large enough that nearby stars will appear to shift their position with respect to more distant, background stars. The apparent shift in position is called *parallax*.

Parallax is easy to demonstrate—simply look at a finger on your outstretched hand with one eye closed, and note its position with respect to the background of the place where you are reading this book (say, a window). Then close your open eye, and open the other. The apparent position of your finger with respect to the background window will have shifted by a startling amount. You can make your finger seem to jump back and forth by rapidly blinking one eye and then the other.

Using one eye and then the other is the equivalent of what astronomers do when they look at the same star from both sides of Earth's orbit around the Sun. The distant, background stars play the role of the window, and the distance between your eyes corresponds to the diameter of Earth's orbit. The angular shift measured for a nearby star can thereby be used to determine the star's distance, which is simply the Earth's orbital diameter divided by the amount of the angular shift. The parallax method, first used by Bessel in 1838 on 61 Cygni, was used to find the distance to Barnard's star in 1916 right after it was shown to be the fastest-moving star on the sky.

The Naval Observatory was interested in the precise positions of nearby stars because in the days before the global positioning system (GPS) satellites and their forerunners, Navy ships had to find their positions at sea by celestial navigation—measuring their position on Earth by working backward from the positions of the stars at night. But many bright stars were close enough to the Sun to move noticeably across the sky, ruining their usefulness for precise navigation unless their movements were measured and annual updates were made. While the Navy thus paid his salary for practical reasons, Harrington was more interested in using the same precise positional information to look for unseen companions to the stars, a thoroughly impractical and considerably more inspired venture.

There were many other stars on the Naval Observatory's parallax program besides Barnard's star. Two of them were named after the Belgian American astronomer who first studied them, George-Achille van Biesbroeck. Van Biesbroeck had fled Europe during World War I and joined Chicago's Yerkes Observatory, where he specialized in finding asteroids, comets, and binary stars. In 1961 van Biesbroeck published a list of 12 stars that were exceptionally faint but also very close to the Sun, meaning that they must be luminosity challenged as stars go—very bright objects will only appear to be faint if they are also very far away. Low-luminosity stars are low-mass stars, and the combination of a low-mass star and one that is very close to the Sun was an irresistable one for Harrington—that is the formula for maximizing the chances for finding an extrasolar planet by van de Kamp's astrometric method, the same reasoning that made Barnard's star so compelling a target.

By 1983, Harrington had been observing two of van Biesbroeck's stars for 10 years, and they were starting to act peculiarly. Neither star was a single star like Barnard's star—van Biesbroeck's eighth star (VB8) was a member of a multiple star system about 21 light years from the Sun, like Barnard's star in the constellation Ophiuchus; while van Biesbroeck's tenth star (VB10) was a member of a binary system lying 19 light years away. Because refracting telescopes distort the light that must pass through their glass lenses, whereas reflecting telescopes completely avoid that problem, the Navy's reflecting telescope had systematic errors (a thousandth of an arcsecond for a year's worth of data) about a factor of 10 times smaller than those of Sproul Observatory's refracting telescope, so Harrington was well positioned to make an important astrometric discovery.

Harrington and his colleagues published a paper in the July 1983 issue of the *Astronomical Journal* that presented evidence for unseen companions orbiting both VB8 and VB10. VB8 seemed to have a companion with an orbital period longer than eight years, while VB10 had a companion with a five-year period. The latter object could have a mass in the range of three to five times that of Jupiter. This mass clearly exceeded Lin and Papaloizou's estimated maximum mass for a giant planet of one Jupiter mass. The *Astronomical Journal* paper carefully referred to the objects as "unseen companions" and did not use the word *planet*. Harrington had found something, but what?

DECEMBER 10, 1984:   Arizona astronomer Donald McCarthy and his supporter, the NSF, issued a press release to announce the likely discovery of the first planet outside the Solar System. The NSF had supported van de Kamp's program at Sproul for decades and dearly wanted to share the credit for McCarthy's more promising detection. The new "planet" was orbiting one of Harrington's stars, VB8.

McCarthy, an astronomer at the University of Arizona's Steward Observatory, and two of his Tucson colleagues had already had their research paper checked over and approved by another astronomer, who acted as the paper's referee. Their paper was already accepted for publication by the *Astrophysical Journal,* and they were ready to go public with their ambitious discovery. Instead of trying to find an extrasolar planet by the laborious and protracted astrometric method of looking for a tiny wobble of the star, the Tucson group had gone to the telescope and effectively taken thousands of pictures of the new planet itself, albeit with a new and powerful technique.

McCarthy used the newly developed technique of speckle interferometry to search the region around both VB8 and VB10 for any faint companions. The search was done at infrared wavelengths (the heat from a warm companion can make it considerably easier for a companion to be seen than at the visual wavelengths our eyes use). Infrared light has a wavelength that can be 10 to 100 times longer than that of visible light. In the infrared, a planet like Jupiter can be a thousand times easier to detect next to a star than at visible wavelengths because of the energy carried off by its own thermal radiation (heat lamps give off most of their energy at infrared wavelengths, and home infrared sensors can detect the body heat from human prowlers at night). The fact that both VB8 and VB10 were intrinsically faint, nearby stars made them prime candidates to be searched for dim companions, because they would not drown out the weak light from their companions as badly as a more massive, brighter star would.

The Earth's atmosphere is a severe problem in taking pictures from the ground sharp enough to reveal an extrasolar planet. McCarthy's speckle technique involved adding together many short exposures of VB8 to try to avoid the blurring that occurs in long exposures and is caused by the Earth's atmosphere. The Earth's upper atmosphere bubbles like a pot on a stove, smearing out any long photographic exposure. Each new bubble bends the light coming from the star in a slightly different direction, so that, seen from the Earth, the star appears to jump around on the sky because of the atmosphere. However, very short exposures (much less than a second in length) are not as blurred, because it takes a fraction of a second for the upper atmosphere to make new bubbles. Very short exposures by themselves are too faint to use, but if you take 10,000 of them and add them together (after adjusting for the shift of the star's location caused by the atmosphere's bubble), then you can get a much better picture. That is just what McCarthy did.

McCarthy and his colleagues estimated that VB8's companion, called VB8 B, was an object with a surface temperature of about 1,700 degrees Centigrade, hot enough to melt iron but cooler than any known star. The team judged the "planet" to be 3 to 10 times as massive as Jupiter, lying at a distance from the faint star just a bit larger than Jupiter's distance from the Sun. The orbital period would then be decades, approximately consistent with Harrington's result of a period greater than eight years. Curiously, even though Harrington had found VB10 to have an even stronger wobble than VB8, McCarthy found no evidence for any companions to VB10.

There ensued a minor brouhaha over VB8's companion, reported in the *Washington Post* a few days later, after Harrington found out about McCarthy's announcement. Harrington pointed out that he and his Naval Observatory colleagues had found VB8 B well before the Tucson astronomers did, and that what had been found was not a planet, but a brown dwarf. McCarthy's team countered that Harrington's astrometric detection, being indirect, did not count—the Tucson group was the first to really "see" VB8 B, and that was what mattered.

A few weeks later, *Science* finally stepped in and settled the affair by proclaiming that what Harrington and McCarthy had jointly found was indeed the first *brown dwarf star* ever discovered. It seemed that brown dwarfs had finally turned from the purely hypothetical to the newly discovered.

A brown dwarf is a star that is too low in mass to become hot enough to start burning hydrogen through thermonuclear fusion, the process that powers hydrogen bombs. Hydrogen nuclei (protons) can only begin to be joined together (or fused) into helium nuclei (containing two protons and two neutrons) when two protons smack each other with incredible energy—the sort of energy that protons have when their temperature is at least 3 million degrees Centigrade. The center of the Sun is five times hotter than that, and the nuclear energy released by the resulting hydrogen fusion is what makes the Sun keep on shining brightly billions of years after it was formed.

The Sun is considerably bigger than most of its neighboring stars, however, many of which are 10 times smaller in mass. Stars with smaller masses begin their lives at lower temperatures, because a star's initial temperature is determined by how much gravitational energy was released in order to make the star. The amount of energy associated with a star's gravitational field depends on its mass raised to the second power (i.e., squared), so a star that is half as massive may have only a quarter of the gravitational energy available to heat it up. Gravitational heating is like the pilot light in a furnace, the small flame that is needed to light the main burner—gravitational heating is needed to light the star's thermonuclear fire. As one considers stars of smaller and smaller mass, then, the amount of gravitational energy eventually gets so small that it can't do the trick anymore; the pilot light effectively blows out, and the thermonuclear fire can never be lit.

Astronomers could calculate that the mass at which the pilot light goes out is about 8 percent of the Sun's mass, or about 80 times the mass of Jupiter. Searches for faint stars had revealed plenty of normal stars with masses down to this level, called M dwarf stars. There was no particular reason to believe that the star formation process should not continue to produce stars with masses below this limit, because sustained thermonuclear fusion only occurs after the star has already been formed, at least for low-mass stars, and thus there is no way for a young star that is still gaining mass to know ahead of time if it is going to make it on Broadway and light itself up.

Brown dwarf stars are the embarassing failures of the star formation process, prospective stars that failed to gain the 80 Jupiter masses of gas needed to ignite their hydrogen fuel. They begin to fade away soon after they are formed, so that by the time they are 10 million years old, they are only about 1 percent as bright as the Sun. By the time they are billions of years old, they have faded so much more that they are extremely difficult to see. Finding more brown dwarf stars became an astronomical obsession after the discovery of VB8 B. It was just the sort of challenge that excites and motivates astronomers—an unseen class of new objects lurking somewhere in the universe, waiting to be found.

Brown dwarf stars were so named by Jill Tarter in her Ph.D. thesis at the University of California, Berkeley, in 1975. Tarter had been prodded by her thesis advisor, Joseph Silk, into studying what role might be played in populating the galaxy by stars that were too low in mass to burn hydrogen and be seen. The likely existence of such objects had been recognized in 1963 by Shiv Kumar of the University of Virginia, who called them black dwarfs. Tarter resisted using Kumar's label, because it was already being used to refer to the final state of a normal star after it had exhausted all of its energy sources. Low-mass stars are called red dwarfs, and so something intermediate between red dwarf and black dwarf seemed appropriate. Tarter came up with the term *brown dwarf,* because these mysterious objects were so poorly understood that no one could predict what color they should be, and there are purists who insist that brown is not a color. The name stuck.

JANUARY 9, 1985:   A few weeks after McCarthy's announcement, Bob Harrington drove over to the Carnegie Institution's DTM, just a few miles away from the Naval Observatory in northwest Washington, D.C., and told Wetherill and me about the status of his astrometric search for unseen companions. He believed that VB8 B really was a brown dwarf and that the perturbation he had seen for VB10 implied the presence of a roughly 5 Jupiter-mass object. The fact that McCarthy had not seen anything around VB10 was encouraging to Harrington, because it meant that it was even more likely that VB10 might harbor a Jupiter-sized planet (smaller than VB8 B) that McCarthy would be unable to see. If that was true, then Harrington would have both the first extrasolar planet *and* the first brown dwarf star to his credit. Harrington, needless to say, was rather pleased with the way things were going.

I was excited to hear Harrington, because I was in the process of trying to calculate the minimum mass of a star, and Harrington's revelations about VB8 B were strong encouragement that I was working on a theoretical problem that had a real-world consequence; the absence of this latter aspect does not normally deter theorists in their choice of problems to solve. The fact that I would derive a theoretical result that was in great danger of either being confirmed or disproven by astronomical observations in the not-so distant future added to my anxiety.

I had become interested in star formation in a roundabout way. My doctoral advisor at UC Santa Barbara (UCSB), Stanton Peale, had asked me to read Safronov's book in preparation for undertaking a Ph.D. thesis on planet formation. Peale was a "celestial mechanic," and I was expected to follow his lead and study the orbital dynamics involved in making planets. I was completely sidetracked, however, by the difficulty that Safronov had chosen to skip over—how to get to the point at which a newly formed star is surrounded by a protoplanetary disk. I took off in the direction of this new challenge, trying to understand how stars and their disks form, with Peale's generous consent and support. Once I had helped figure out how stars form, I thought, I would get back to investigating planet formation.

My thesis project was enabled by the same development that allowed Wetherill to go far beyond Safronov's work on planetary accumulation—the arrival of digital computers, such as the IBM 360/75 mainframe at UCSB. I undertook the lengthy task of writing from scratch a computer code that would predict how an interstellar cloud of gas and dust would collapse to form stars. The code was designed to follow the contraction and distortion of interstellar clouds in all three dimensions of space, a problem much too difficult to be solved without a computer. But running a three-dimensional computer code required a fast, big computer, and soon I was flying to Los Alamos, New Mexico on the weekends to use the Control Data Corporation (CDC) 7600 computers, while the resident nuclear weapons designers went skiing and hiking in the surrounding mountains, leaving the Los Alamos computers idle. I used my new code to study the problem of binary star formation, publishing with Peter Bodenheimer of UC Santa Cruz (UCSC) in 1979 the first comparison of two different computer calculations showing how an interstellar cloud could collapse and fragment into two clumps that might later form a binary.

After finishing my degree in 1979, I joined David Black at NASA Ames as his postdoctoral fellow. Black had become interested in interstellar cloud collapse in the mid-1970s and published a pioneering study of the collapse of rotating clouds with Bodenheimer. I spent the next two years at Ames continuing my calculations of binary formation, working mostly by myself—Black had become heavily involved in helping to manage NASA's affairs and had little time left for research. Like van de Kamp, Black had developed a fondness for sleek cars, the kind of car that people often park across the middle of two parking spaces to prevent accidental damage.

At Ames I made a key improvement to the basic computer code, something I had been planning to do for years. I gave my code the ability to calculate not just the motions of gas and dust in clouds, but also the flow of radiation, which controls the temperature of the cloud. By the time I left NASA Ames in 1981, the new code was ready to run, but I had run out of time and was unable to do more than run a few tests.

In the fall of 1981, I joined George Wetherill and the eclectic staff of the DTM. DTM's permanent research staff was composed of three distinct groups: extragalactic astronomers, seismologists, and geochemists. The research interests of the three groups matched those of an earlier, highly influential DTM Director, Merle Tuve, more than they matched what would be expected for a department seemingly devoted to the study of the Earth's magnetic field. In fact, terrestrial magnetism was a topic that was definitely *not* being actively worked on at DTM. Wetherill was the only person working on Solar System formation, so my arrival would double DTM's small effort in a field that represented yet another new direction for the Department.

FIGURE 12. Peter Bodenheimer, professor of astronomy at the University of California, Santa Cruz and a leader in studying the hydrodynamics of the formation of stars and gas giant planets. (Courtesy of P. Bodenheimer.)

Before accepting Wetherill's offer, I had asked Stuart Weidenschilling over lunch at Ames what DTM was like. Weidenschilling had spent two years at DTM as a postdoc, and he summed it up as being "genteel." Genteel it was, and a bit antiquated and dusty to boot. The older buildings were largely unchanged from when they were built well over half a century earlier. In fact, DTM did not have a computer capable of handling my codes, so my star formation research was put on hold for over a year. By 1983, DTM and another nearby Carnegie department, the Geophysical Laboratory, had acquired two Digital Equipment Corporation VAX-11 computers, and I was back in business again. Soon Wetherill was busy making Earth-like planets on the VAX computers, and I was finally ready to try the new code I had developed at Ames.

The problem I intended to attack was the question of the minimum mass that a star could attain. James Jeans, of tidal hypothesis fame, had derived in 1928 a criterion for how much mass it takes to make a star out of an interstellar cloud. Though highly idealized, Jeans's analysis had stood the test of time as a handy criterion, and the concept of the "Jeans mass" had become part of every astronomy graduate student's education. An interstellar cloud can be held back from falling in on itself by the pressure of its gas, and Jeans calculated how big a cloud would have to be for its own gravity to overwhelm its thermal pressure and lead to its collapse. A large interstellar cloud contains enough gas to make thousands or even millions of stars.

Once a portion of an interstellar cloud begins to collapse in upon itself, however, it can break up into even smaller pieces. A cloud's gravitational attraction increases greatly as the cloud's density increases, faster than the gas pressure increases, at least at first. The Jeans mass thus decreases with increasing density, allowing smaller and smaller mass clumps to form. This process is called *fragmentation*. Fragmentation is how most binary stars form, as Bodenheimer and I had crudely hinted in 1979. But the fragmentation process need not occur just once or stop at masses like that of our Sun—a hierarchy of fragmentation could ensue, leading to progressively smaller clumps. This idea had been advanced in 1953 by the imaginative British astrophysicist and science fiction writer Fred Hoyle, who also took delight in spinning theories about the formation of the universe.

Eventually, though, fragmentation has to stop. This occurs once a cloud becomes so dense that it starts to heat up—the Jeans mass starts to get larger as the temperature of the cloud increases, and then fragmentation ceases in this simplified picture. The smallest value reached by the Jeans mass is then a good estimate of the smallest mass clump of gas that can pull itself together into a star (i.e., the minimum mass of a star). This minimum mass was estimated to be about 7 Jupiter masses by Cambridge astrophysicists C. Low and Donald Lynden-Bell in 1976. Their colleague Martin Rees almost simultaneously came up with an even lower value, about 2 Jupiter masses. Thus there appeared to be a range of masses within which brown dwarf stars could happily romp without much fear of being seen as hydrogen-burning stars—they could have masses anywhere from about 2 to 80 times that of Jupiter.

However, a Jeans mass analysis is only expected to be approximately accurate at best, because it is based on a highly simplified model that neglects most of the messy details of star formation and fragmentation, such as the fact that clouds rotate and contain magnetic fields. A year later, another theoretical astrophysicist, Berkeley's Silk, also used the Jeans mass approach to calculate the minimum mass of a star but came up with values ranging from 10 to 100 Jupiter masses or more when some of these other effects were factored into the analysis. If the latter value was accurate, then there could be no brown dwarfs at all—*all* stars would be massive enough to burn hydrogen.

The stakes suddenly were high. If Silk's higher estimate was right, then brown dwarfs simply did not exist. If the Cambridge estimates were right, then brown dwarfs could exist, but some might have masses close to the mass of the most massive planet known at the time, Jupiter. Being composed largely of hydrogen and helium gas, brown dwarf stars have a physical structure similar to giant planets like Jupiter, so if there was no significant gap between the mass of the least massive brown dwarf and the most massive planet, then there was sure to be great confusion in telling one from the other once something was found in that mass range.

A more intensive study of this question was clearly warranted. The new code I had developed at NASA Ames was perfectly suited to study the problem because it could calculate the processes of heating and cooling that control the Jeans mass and determine when fragmentation must stop. In addition, the code could treat many of the complications that were certain to arise in real clouds, such as three-dimensional distortions and rotation, that could not be included properly in the simple Jeans mass analysis. Because it is particularly easy to handle calculations for a sphere, in which quantities vary only in radius, theoreticians frequently use a sphere to represent any astronomical object that they wish to study and are often chided for this tendency. If they were suddenly asked to study veterinary medicine instead, some theorists might start an analysis by saying, "Consider a spherical cow . . ."

With the right computer code, on the other hand, the restriction to spheres could be dropped, and the fragmenting clouds could be shaped like four-legged cows, if that was what was indicated by observations. I started my calculations in 1983, and by 1985 I had finished enough runs to be able to make a prediction. Harrington's and McCarthy's discovery gave added urgency to the work, along with a major conference that was about to occur.

OCTOBER 14–15, 1985: The excitement of the discovery of VB8 B prompted the first conference ever devoted to the subject of brown dwarf stars, held at George Mason University in the Virginia suburbs of Washington, D.C. Bob Harrington and Don McCarthy were the cochairs and had the honor of giving the first two talks. Kaj Strand sat quietly in the audience, his 1943 claims of a "planet" around 61 Cygni long forgotten and unspoken, though his Naval Observatory colleague Harrington did speak favorably about Strand's later work on another star. Harrington also mentioned that his 12 year study of Barnard's star showed no sign of a planetary companion. Van de Kamp did not attend the conference.

The George Mason conference made it clear that brown dwarfs were suddenly a hot topic. A dozen astronomers said they were busily looking for brown dwarf stars with a bewildering variety of methods, but they had all come up empty handed. VB8 B was the only star of the show and had the spotlight all to itself.

Brown dwarf theorists had been busy making models for how fast a brown dwarf of a given mass would cool off with time. From the estimated luminosity of VB8 B, they could then tell how massive VB8 B must be, if it was about the same age as the Sun, as good a guess as any for the denizens of the Sun's neighborhood. On this basis VB8 B seemed to contain at least 50 Jupiter masses, considerably higher than Harrington's latest astrometric estimate of 20 Jupiter masses, but well within the expected range for very-low-mass stars that would never burn brightly like the Sun.

Late on the second and final day of the conference, I presented the results of my new calculations. My computer models showed that the minimum stellar mass should be about 20 to 50 Jupiter masses, just below the range suggested by the previous speakers for the mass of VB8 B. For the moment, at least, theory and observation agreed. In addition, David Black and I each pointed out that anything smaller than that would have to form like a planet, out of a protoplanetary disk, rather than as a star, by the collapse of a cloud.

The conference broke up with a feeling of general satisfaction and anticipation for the future. Brown dwarfs had been given the stamp of approval by the relevant astronomical community, and the Cambridge University Press would publish the first book about brown dwarfs and the George Mason conference the following year.

MARCH 23, 1986: Two French astronomers, C. Perrier and J.M. Mariotti, went to a mountaintop in Chile attempting to find VB8 B by the same speckle interferometry method used by the Tucson astronomers. They had the use of a large telescope equipped with a new "infrared specklegraph," and the weather was cooperative—but they could not find the brown dwarf companion to VB8. They repeated their observations several times during the night but still found no trace of VB8 B. In a paper submitted to the *Astrophysical Journal* a few months later, they ruefully reported their multiple failures to confirm the Tucson discovery, and, worst of all, suggested that VB8 B might not even exist—the Tucson team may have been misled by the insidious effects of the Earth's atmosphere. The arrival of the French manuscript at the editorial office of the *Astrophysical Journal* in Tucson, located just across the street from the Steward Observatory, must have caused severe consternation for the Tucson group.

The French were not the only ones who could not find VB8 B. Michael Skrutskie of Cornell University and two colleagues had used a unique telescope, specialized to work at infrared wavelengths, to search for VB8 B even before the French did. On the night of July 28, 1985, they had used NASA's Infrared Telescope Facility (IRTF) on Mauna Kea, Hawaii to look for VB8 B during a period when the atmosphere was extraordinarily calm and permissive of sharply focused photographs. They used a camera designed by one of the team members, William Forrest of the University of Rochester, New York, optimized to work at infrared wavelengths. They had reported at the George Mason meeting their failure to detect any brown dwarfs around the 68 stars they had observed in July 1985, but the group coyly had not mentioned that their nondetections included VB8 B. Almost one year later, they had analyzed their data as best they could and still could not find VB8 B. They sent their own paper to the *Astrophysical Journal* on September 2, 1986. That made two successive failures to find VB8 B again.

The Cambridge University Press book was still in production when VB8 B began to disappear as a result of the work by the French and New York state astronomers. McCarthy admitted that it was possible that VB8 B was a chimera caused by bad luck, but suggested that perhaps the other teams could not find VB8 B simply because it had moved closer to its star VB8 in the time interval between his measurements and theirs, and that was why VB8 B seemed to disappear. If so, the elusive brown dwarf would soon enough reappear once again, and then everyone would be satisfied.

VB8 B never did reappear. The infant field of brown dwarf research, born with great promise just a few years before in 1983, seemed to be in danger of an untimely death.

6

WE KNOW YOU ARE OUT THERE...

*Research means going out into the unknown with the hope of
finding something new to bring home. If you know in advance
what you are going to do, or even to find there, then it is not research at all: then it is only a kind of honorable occupation.*

—Albert Szent-Györgyi (1893–1986)

The overall situation looked pretty grim in late 1986. With the "first
brown dwarf" VB8 B having turned into an excessively shy recluse,
Harrington's astrometric hopes for the "first planet" VB10 B also
began to fade. The astrometric claims for VB10 B would soon be politely forgotten.

Van de Kamp still believed that he had evidence for planets circling
Barnard's star. In 1986 he had published his last book, *Dark Companions of Stars,* detailing his efforts through the decades. In a telling manner, the frontispiece for the book consisted of a Rembrandt etching of a
biblical scene, with a quotation from the Bible—"Blessed are they that
have not seen, and yet have believed." The quotation perfectly summed
up van de Kamp's philosophy about his life's work on finding unseen
companions to stars. Van de Kamp believed that he could only be vin-
dicated about Barnard's star when someone else had gathered many
decades of data, like he had.

Harrington finally published the results of his own work on
Barnard's star in early 1987 in a popular article he wrote for the astronomy magazine *Mercury.* Harrington's results gave no support for
van de Kamp's claims, even though Harrington had analyzed 14 years
of data from the Naval Observatory's much superior telescope—
Barnard's star was still planetless. Harrington did not mention the re-

lated but exceedingly sensitive topics of VB8 B and VB10 B in his Barnard's star article.

Patience was truly a virtue in van de Kamp's world, and van de Kamp was willing to wait. But the rest of the world wanted planets and brown dwarfs, and wanted them *now*.

While much of the general expectation that planetary systems should be widespread could be traced to the nebular hypothesis, clear astronomical evidence was beginning to emerge that stars were often surrounded by dense disks of gas and dust that would be perfect places for building planets. One major advance was made by the *Infrared Astronomical Satellite (IRAS)*, built by NASA and its Dutch and British counterparts and launched into Earth orbit in 1983. By virtue of its location well above the Earth's watery atmosphere, the *IRAS* telescope was able to survey the sky at infrared wavelengths between 12 and 100 microns, its measurements free of contamination by molecules in the atmosphere.

IRAS found convincing proof that stars do indeed form out of dense interstellar clouds. For example, Charles Beichman of NASA's Jet Propulsion Laboratory (JPL) in California and his colleagues used *IRAS* to examine the dark cloud Barnard 5, the fifth cloud in a listing of dark clouds compiled by the same Edward Barnard who found Barnard's star. Beichman and his team found in 1984 that Barnard 5 contains two sources of infrared radiation—most likely young stars still in the process of formation. To our eyes, Barnard 5 appears as an irregular, black region on the starry sky, because its dust grains block out the light coming from stars unfortunate enough to lie behind it. But to the infrared eye of *IRAS*, such dark clouds are spotted with tiny but bright sources of infrared light—just what would be expected if young stars were buried deep within the clouds out of which they were still forming. The young stars would emit most of their light at visible wavelengths, like the Sun, but this light would be absorbed by cool dust grains surrounding the stars and then emitted again at infrared wavelengths, making the young stars appear to be infrared stars.

Other young stars are not so deeply hidden in clouds that they could not be seen by the eye using a telescope, but even these optically visible young stars were seen by *IRAS*. Young stars have surface temperatures of thousands of degrees and give off most of their energy at visual wavelengths. *IRAS* found that many of these young stars also gave off much more radiation at infrared wavelengths than they should, based on their surface temperatures. These "infrared excesses" could only mean one thing—there had to be a large number of dust grains in orbit around these stars that could emit the infrared light. The dust grains could not be part of a dense spherical shell entirely surrounding the young star, because then they would hide the star from our view. Instead, the grains had to lie in a more restricted region of space around the star (e.g., in a disk). The *IRAS* measurements were soon shown to be in basic agreement with the predictions made by Lynden-Bell and Pringle in 1974 for the light given off by a viscous accretion disk. These *IRAS* observations thus pointed to the likely existence of protoplanetary disks orbiting young stars, though the evidence was indirect by nature.

Both of these types of *IRAS* observations helped to prove that the Kant-Laplace nebular hypothesis was correct, not that there was any remaining doubt. Accordingly, planets had to be common, even if we could not yet see them. But there was another, more surprising discovery to come as a result of *IRAS*'s work.

There was something strange going on with the nearby bright star Alpha Lyrae, as well as with a handful of other stars observed by *IRAS*. The rumors began to spread soon after *IRAS* was launched in January 1983. Alpha Lyrae, also known as Vega, is over three times as massive as the Sun and is the fifth brightest star in the sky. Alpha Lyrae was a natural choice to be used to calibrate the *IRAS* instruments. Because astronomers thought they knew exactly how much infrared light Alpha Lyrae was giving off, they would simply adjust the instruments on *IRAS* so that the "right" value was obtained for Alpha Lyrae, similar to setting the zero point on an empty scale before weighing yourself. But Alpha Lyrae, and a few other stars, gave off far more infrared light than could be explained by the expected small uncertainties in the readings from the *IRAS* instruments; it was as if something heavy was standing on the scale, even though nothing was supposed to be there. Clearly Alpha Lyrae had an infrared excess similar to that found for young stars. JPL's H. H. Aumann and his colleagues concluded that Alpha Lyrae was surrounded by a spherical shell of dust grains left over from its formation no more than 300 million years earlier.

Bradford A. Smith of the University of Arizona was tipped off by a friend on the *IRAS* team about the excess infrared light from the *IRAS* stars, and Smith was in a position to do something about the hot tip. He had just developed a device, called a coronagraph, that could be mounted on a telescope and used to search for faint rings around planets like Uranus and Neptune, ethereal rings that are the poor cousins of Saturn's mighty ring system. The coronagraph had a central disk that blocked out most of the light from a bright object in the center of the telescope's field of view, permitting fainter objects to be seen in the vicinity of the bright object. The coronagraph was the perfect instrument to use to look for anything peculiar in orbit around the strange *IRAS* stars.

Smith immediately used the coronagraph to photograph the region around the *IRAS* stars Alpha Lyrae and Epsilon Eridani, both easily visible from the mountaintop observatories that encircle Tucson. In 1974, van de Kamp suspected that Epsilon Eridani had a 6 Jupiter-mass companion, and while no confirmation had been forthcoming, Epsilon Eridani was a tempting target given the *IRAS* measurements. But Smith came up empty handed: His coronagraph found no evidence for anything orbiting either star. Smith also tried a few other stars similar in mass to Alpha Lyrae and Epsilon Eridani, but still found nothing. But Smith did not give up at that point—perhaps he had picked up some of Clyde Tombaugh's gritty perseverance while working for Tombaugh before entering graduate school.

Another one of the peculiar stars was Beta Pictoris. Beta Pictoris is a normal star like the Sun, though twice as massive, located 53 light years away on the outskirts of the Sun's neighborhood in a constellation that cannot be seen from Tucson. Smith would have to wait until he had an observing run scheduled in the southern hemisphere before he would be able to examine this star with his coronagraph. Because Beta Pictoris was known to be losing mass through a stellar wind, the expectation was that Beta Pictoris was surrounded by a thin, spherical shell of dust grains. Maybe Smith would be able to detect this shell of dust grains blown off of the star.

Luckily, Smith and JPL's Richard J. Terrile had an observing trip to look for planetary rings scheduled for early in 1984, and they took Smith's coronagraph along with them to the Carnegie Institution of Washington's Las Campanas observatory, located in the foothills of the Chilean Andes mountains. Each night during April 15–18, 1984, they spent an hour observing Beta Pictoris with their coronagraph, but they could not tell if they had found anything from looking at their raw data. It wasn't until July that Smith and Terrile got around to analyzing their Beta Pictoris images. They were shocked when they immediately found two "bright spikes" sticking out on either side of the blotted-out position of Beta Pictoris—instead of a spherical shell, Beta Pictoris had a *disk* that could be seen with visible light. The disk extended for a distance at least 400 times greater than Earth's distance from the Sun, 10 times beyond the orbits of the planets of our Solar System.

The disk could only be attributed to a huge number of dust grains orbiting around Beta Pictoris. The dust grains could not be leftovers from the formation phase of Beta Pictoris, because such small dust grains could not last long in the presence of the radiation field of the star. Evidently, Beta Pictoris harbored a population of larger objects, perhaps comet-sized bodies, that somehow collided and replenished the dust grains in the disk.

The discovery of the Beta Pictoris dust disk provided the first unambiguous, direct proof that flattened disks of matter exist around stars other than the Sun. Smith and Terrile's discovery was celebrated by the media with stories in *Science* and the *New York Times* and elsewhere. As usual, the nebular hypothesis ruled.

The next major observational discovery about protoplanetary disks came from radio telescopes designed to work in the millimeter range (one inch contains 25.4 millimeters). Very cold dust grains, with temperatures of hundreds of degrees Centigrade below zero, give off radiation in the form of millimeter-length radio waves.

Millimeter-wave radiation is a form of light, just like the visible light our eyes sense, but the wavelengths are thousands of times longer. Millimeter-wave radiation from astronomical objects can be detected by building reflecting telescopes similar to optical (visible light) telescopes, except that the surfaces of millimeter-wave telescopes consist of polished metal rather than aluminum-coated glass, in order to reflect the millimeter waves to the telescope's focal plane more efficiently. Rather than having photographic plates, millimeter-wave telescopes have electronic detectors that measure the incoming radiation. These days, most optical telescopes also have electronic devices to collect the light, because such devices waste less light than a photographic plate does and because they yield computer-readable data in a single step.

Millimeter-wave telescopes must be much larger than optical telescopes in order to see objects with as fine a detail because of the much longer wavelengths of the light they detect. In fact, in order to have the same resolving power as an optical telescope, a millimeter-wave telescope needs to be over a thousand times larger in size.

The world's largest fully functional telescope in the mid-1980s was still the 1948-vintage Hale Telescope on Palomar Mountain in southern California, with a primary reflecting mirror 200 inches in diameter. To match its resolution, a millimeter-wave telescope would have to have a dish over 3 miles in diameter, a ridiculously impossible goal. But millimeter-wave astronomers could use a clever method to gain most of the advantages of a supersized telescope without the supersized cost: They could link several smaller telescopes together into an *interferometer* and use a high-speed computer to combine the individual telescopes into what is, in effect, a single gigantic dish. An interferometer gets its name from the need to combine the radio waves coming from each telescope so that they constructively interfere, the same thing that happens when two ocean waves momentarily merge together to make an even larger wave.

With the support of the NSF, Caltech astronomers designed a millimeter-wave interferometer in the Owens Valley, just east of the Sierra Nevada mountains in California. By 1984, they had built three telescopes over 30 feet in diameter and were able to separate the telescopes by distances up to 200 feet and still combine the millimeter-wave signals in much the same way as if they had been able to build a single, monolithic telescope 200 feet in diameter.

With this powerful new tool, Steven V. W. Beckwith of Cornell University and Caltech's Anneila I. Sargent set out to take the first picture of a protoplanetary disk.

MARCH 5, 1987:   After several years of efforts, Sargent and Beckwith submitted a paper to the *Astrophysical Journal* with a striking map of a disk around the young star HL Tauri, a variable star in the Taurus constellation. Their first attempt, published just the year before, had shown little more than a hint of an elongated region of millimeter-wave emission from HL Tauri. Those measurements had been taken during the spring and summer months, when water vapor in the Earth's atmosphere can be a problem for observations at millimeter wavelengths. Sargent and Beckwith's new observations were taken in the winter months, with an otherwise identical set up for the interferometer, and the results were vastly better.

Sargent and Beckwith chose HL Tauri for their ambitious attempt because it was known to be a young star of the same general type (the T Tauri stars) that are believed to turn into stars like the Sun. Hence it was a good bet to be an analog for the protosun and solar nebula. HL Tauri also happened to be fairly close to the Sun, lying in the Taurus star-forming region just about 450 light years away. Because Taurus is one of the closest star-forming regions, it has become a favorite stomping ground for astronomers seeking to discover new aspects of the star formation process. With such a nearby young star, Sargent and Beckwith had a good chance of being able to see something with the Owens Valley interferometer, given its maximum effective size of about 200 feet (equivalent in resolving power, however, to an optical telescope only a few inches in diameter).

The interferometer could be used to observe HL Tauri simultaneously in two ranges of wavelength, so that twice as much information

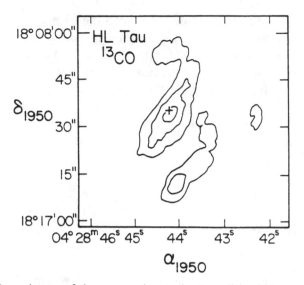

FIGURE 13. An early map of the suspected protoplanetary disk orbiting around the young star HL Tauri (plus sign). The flattened disk was imaged at millimeter wavelengths through emission from ^{13}CO molecules. (Reprinted, by permission, from A. Sargent and S. Beckwith, 1987, *Astrophysical Journal,* volume 323, page 295. Copyright 1987 by the American Astronomical Society.)

was obtained during the several months of observations. One band looked for emission from dust grains around HL Tauri, while the other looked for radiation from molecules of carbon monoxide gas, the poisonous gas commonly found in automobile exhaust and interstellar clouds.

The carbon monoxide emission provided the more striking image. Sargent and Beckwith found that HL Tauri was encircled by an enormous flattened disk of molecular gas. The disk seemed to have about 10 percent as much mass as the Sun, extending 50 times farther away from HL Tauri than the orbit of the most distant planet (Pluto) in our Solar System. To paraphrase Crocodile Dundee, "Now *this* is a *disk*."

The *IRAS* infrared protostars and disk infrared excesses, the Beta Pictoris disk, and the HL Tauri disk all combined to make a convincing proof for the theorists' basic concepts of how stars and planets form. Extrasolar planets just *had* to be out there. Most likely, so did brown dwarf stars. But who would find the first real ones?

NOVEMBER 12, 1987: Benjamin Zuckerman of UCLA and Eric Becklin of the University of Hawaii published a paper in *Nature* detailing their observations of the infrared energy being emitted by the white dwarf star Giclas 29-38. Using the IRTF on Mauna Kea, they found that this white dwarf star had an excess of infrared emission, just like *IRAS* had found to be the case for many young stars. As far as they could tell, the extra radiation did not come from some background object that just happened to lie in the same direction as Giclas 29-38. If the excess infrared light was really coming from Giclas 29-38, once again this meant that something warm must be in orbit around the hot star. It could be a swarm of dust, or it could be something much bigger.

Zuckerman and Becklin decided that what they had bagged was a brown dwarf star orbiting Giclas 29-38. A swarm of dust would have a short lifetime around Giclas 29-38, only a decade or so, and so would have to be replenished by some larger body in orbit, which seemed unlikely to them. If you had to invoke a larger body in orbit anyway, why not go all the way? A brown dwarf star could orbit around Giclas 29-38 indefinitely, and if the brown dwarf had a surface temperature of about 900 degrees Centigrade, it would be expected to give off enough infrared emission to match the IRTF observations. Zuckerman and Becklin estimated that a brown dwarf star with a mass between 40 and 80 times that of Jupiter would do.

White dwarf stars are one of the three types of exotic outcomes (the others being neutron stars and black holes) that result from stars when their normal lives are over (i.e., after they have fused as much hydrogen into helium as they can). White dwarf stars are so named because they are exceedingly hot stars (white hot, in fact) but with diameters comparable to that of the Earth, in spite of having about as much mass as the Sun. Because of their small diameters, white dwarf stars are much harder to see than other stars with such hot surfaces. The unseen companion to wobbly Sirius, found by Bessel in 1844, was one of the first white dwarf stars ever found.

But next to a white dwarf star was not exactly an expected place to find a brown dwarf star. The Sun will turn into a white dwarf star in another 5 billion years or so, but only after it first swells greatly in size during its red giant phase of evolution. The outer layers of the Sun may then expand out to the orbit of the Earth, enveloping it and extinguishing any life forms that might still be around at the time. The red giant star loses much of its mass through a strong wind and then contracts down to become a white dwarf star. If a preexisting planet or brown dwarf star was orbiting close enough to be swallowed by the red giant star's outer layers, it might well undergo some understandably radical changes.

NOVEMBER 16, 1987: The *Washington Post* hailed Zuckerman and Becklin's discovery of a brown dwarf companion to Giclas 29-38 as the first "planet-like object" found outside the Solar System. The discovery was presented as evidence for the likely existence of many other planetary systems in the Sun's neighborhood. The article presented the difference between stars and planets as being simply one of mass, disregarding the theoretical belief that stars and planets form at different phases and through different mechanisms during the overall process of star and planet formation. *Science* published a more restrained article two months later, emphasizing that it was a brown dwarf that had been found by Zuckerman and Becklin, not a planet.

The next step was for somebody to confirm Zuckerman and Becklin's discovery, in order to make sure that a brown dwarf had been found at last. Alan Tokunaga's University of Hawaii group, including Becklin and other collaborators, had already used an IRTF spectrometer in September 1987 to study in more detail how much radiation was being emitted by Giclas 29-38 and its companion at different wavelengths. They hoped to find some evidence for decreased emission around 2 microns in wavelength that would be indicative of the presence of atoms or molecules in the relatively cool atmosphere of the brown dwarf.

They found none. Tokunaga and his group also tried imaging the region around Giclas 29-38 with an infrared camera on a University of Hawaii telescope also on Mauna Kea, but found no evidence for anything other than Giclas 29-38 itself. Evidently, this brown dwarf was not going to be easy to confirm.

Another team used the same infrared speckle interferometry technique that had come to grief on VB8 B to look for Giclas 29-38 B. At first they thought they found a hint of something in orbit, and they submitted a paper to be published in the European journal *Astronomy and Astrophysics;* but before the result could be published, they failed to find the same effect in their observing run the next year.

A group from Caltech, the University of Texas, and Columbia University put forward evidence in 1990 that the infrared excess of Giclas 29-38 came from a dust cloud, not a brown dwarf star. James Graham and his colleagues had found that the infrared emission pulsed with periods identical to that of optical light coming from the white dwarf Giclas 29-38, whose light output increases and decreases slightly with several periods in the range of a few to ten minutes. Graham and his team suggested that the fact that the infrared emission pulsed right along with the white dwarf star meant that the emission probably came from a dust disk in orbit around Giclas 29-38, rather than from a brown dwarf star—the dust disk would be periodically heated and cooled by the radiation from the white dwarf star and would respond by giving off pulsed infrared radiation. Graham's group also found that with refined measurements, the infrared excess of Giclas 29-38 could no longer be explained by a single brown dwarf star, with its single surface temperature, but a dust disk with a range of temperatures could explain the infrared observations.

By 1990, Giclas 29-38 B had metamorphosed from a brown dwarf into another disk of warm dust, an interesting phenomenon, but not what we needed. Another brown dwarf had slipped away into the depths of the night sky.

If we could not find brown dwarf stars, how would we ever find the even fainter extrasolar planets?

7

...And We Are Coming
to Get You

First, I believe that this nation should commit itself to achieving the goal, before this decade is out, of landing a man on the moon and returning him safely to earth. No single space project in this period will be more exciting, or more impressive to mankind, or more important for the long-range exploration of space; and none will be so difficult or expensive to accomplish.

—John Fitzgerald Kennedy (1917–1963), May 25, 1961

It was becoming clear in the 1980s that business as usual would not do. Astronomers were not going to be able to find Jupiter-like extrasolar planets and brown dwarf stars, much less Earth twins, simply by using existing instruments and telescopes, devices that were not designed with the delicate task of extrasolar planet detection in mind. Extrasolar planets and brown dwarfs were turning out to be so hard to find that the billionaire recluse Howard Hughes looked like a shameless publicity hound in comparison. It looked like the challenge of finding these overly shy extrasolar beasts would require building special-purpose instruments and telescopes. That meant big money was needed, and that meant it was time for Uncle Sam to get involved in a more proactive manner.

Some of the ideas that were floating around would cost really big bucks. In a 1978 letter to *Nature,* Stanford University electrical engineer Ronald N. Bracewell had proposed building an interferometer in Earth orbit that would be used to look for extrasolar Jupiters. The interferometer would work at infrared wavelengths, where it could take advantage of the enhanced radiation coming from warm planets to enable the *direct* detection of extrasolar planets—instead of settling for finding an astrometric wobble of the star and then inferring the presence of a planet, the infrared interferometer would detect light actually coming from the planet itself. To avoid the infrared emission of the Earth's atmosphere, however, the interferometer would have to go into space. Such an ambitious telescope, like anything that goes into space, would cost considerably more than a similar-sized ground-based telescope. The financial support of the federal government would be necessary to design, build, launch, and operate the interferometer—there was no way that a few astronomers could assemble the resources needed to bring off such a major enterprise by themselves.

By virtue of being placed in space, the development of the space infrared interferometer and other similar ideas naturally fell within NASA's bailiwick rather than within NSF's. With a few notable exceptions, such as NASA's Infrared Telescope Facility in Hawaii, ground-based telescopes were supported only by the NSF, while NASA had dominion over anything, as the joke goes, that could make it more than an inch above the ground, whether it was a telescope on an airplane, on a balloon, or on an Earth-orbiting satellite. With the NSF already being the big player in the world of U.S. astronomy, but with NASA beginning to show definite signs of serious interest (e.g., *IRAS*), this was a natural way to divide the turf and to avoid interagency fighting over sponsorship of future projects.

Stanford's Bracewell had been spurred into conceiving his novel idea in part by attending workshops held at NASA's nearby Ames Research Center. Bracewell had a long-standing interest in trying to communicate with intelligent beings on other planets—he had written a popular book about the subject in 1974. When scientists at NASA Ames began holding workshops about mounting a serious search for extraterrestrial intelligence (SETI) in 1975, Bracewell was invited to become a member of the group. He was thus well motivated to think of a new way of finding extrasolar Earths, though in his *Nature* letter he was content to imagine being able to find extrasolar Jupiters, a much simpler task but one that had stubbornly managed to elude solution.

Directly following the Ames SETI workshops in 1975 and 1976, a three-month workshop was held also at Ames that dealt not with searches for extraterrestrial life, but with designing a system to search for extraterrestrial planets. Part of the justification for the effort was to find likely targets for the SETI project. The SETI radio telescopes would be attempting to discover if another civilization was intentionally broadcasting its existence to the galaxy, and it would help immensely to know ahead of time which stars had planetary systems worthy of telescope time and which did not.

The 1976 summer workshop was organized by David Black, whose interest in finding extrasolar planets had been whetted in 1972 by the heady hopes for Barnard's star. Bracewell and George Gatewood served as advisors for the 1976 workshop, which focused primarily on figuring out how to build a ground-based interferometer that could be used to detect extrasolar planets indirectly through van de Kamp's astrometric technique. The design produced by the workshop's participants involved twin optical interferometers, each 50 meters in length, that would measure stellar positions a factor of 30 times better than existing telescopes. The participants went into great detail in designing the imaginary telescope, going so far as to specify how thick the concrete foundations should be for the telescope supports (3 feet thick). Oddly enough, because of its interferometric design, the proposed telescope would not be able to look for the wobble of the closest stars, for which the wobble would be most pronounced. This was because the powerful interferometer would be confused by its ability to detect surface features on the closest stars, such as sunspots, so that the measurements of the star's astrometric wobble would be ruined. The interferometric technique demanded that the target stars be distant enough to be mere blurs.

In 1980, Black published the results of this Project Orion, jauntily named after the constellation of Orion the Hunter, which contains the Orion molecular cloud, a well-studied region of active star formation. The Project Orion report called for NASA to support a program of searching for extrasolar planets and to continue to study the perplexing technical aspects of the problem of how best to find planets outside our Solar System.

The National Academy of Sciences had issued its own report in 1969, calling for the United States to build a "Large Space Telescope" that, freed of the Earth's atmosphere, would do wonders for astronomy as a whole and might even be able to detect newly formed protoplanets. In an Appendix to the 1980 Project Orion report, Black's team stated that the Space Telescope, by then being developed by NASA, should be able to detect Jupiters orbiting Sun-like stars as far as 33 light years away. Assuming perfect optics, the Space Telescope would be able to find a planet directly, simply by taking a 20-minute exposure with its electronic cameras. The Project Orion proposal involved using the Earth's Moon to block the light coming from the planet's star. The

Moon essentially would be used as a colossal "occulting disk," like the small one in the coronagraph that Smith and Terrile used to see the disk around Beta Pictoris. The technique would only work for a few dozen bright, nearby stars, however, and these by and large did not happen to lie along the path of the Moon on the sky, where they would eventually pass behind the Moon as seen from the Earth. Hence the Project Orion group suggested putting the Space Telescope not in the expected low Earth orbit, just a few hundred miles above the Earth's surface, but rather on an orbit similar to that of the Moon, which orbits at a distance of 30 times the diameter of the Earth. In this Moon-like orbit, the Space Telescope could maneuver around and use the Moon to block out the light from target stars all over the sky. The Project Orion proposal had no chance of succeeding, however, because the astronomers who were building the Space Telescope were not about to recommend changing their plans and sending their precious creation off to the wilds of the Moon's orbit to hunt for planets that may or may not even be there.

The Project Orion report sailed out into a largely unreceptive world and disappeared from sight. The 1980 call for a national program to find extrasolar planets was a bit ahead of its time. The field of extrasolar planet detections did not as yet have a single success story to point to as uncontested evidence of the reality of these still hypothetical systems.

But the issue would not go away. In December 1985, the Solar System Exploration Division of NASA created an ad hoc Planetary Astronomy Committee and charged it with the task of thinking up new projects to undertake. With David Black as a member, perhaps not surprisingly the Planetary Astronomy Committee singled out the largely untouched area of looking for extrasolar planets as an endeavor worthy of the effort. The committee specifically called for NASA to build a new astrometric telescope, to be flown in space as a part of NASA's planned Space Station. Black had just moved from Ames to NASA's headquarters in Washington, D.C., where he had become the Chief Scientist for the Space Station then being planned. To help justify its tremendous cost, the Space Station needed as much scientific work to do as possible, and, from a scientist's focused point of view, the search for extrasolar planets seemed like a creative way to give the Space Station's astronauts something important to do besides wait to get back down to Earth.

The space astrometric telescope proposed by the Planetary Astronomy Committee was expected to be able to measure stellar wobbles as small as 10 microarcseconds (one degree of angle is divided into 3,600 arcseconds, and a microarcsecond is a million times smaller yet), a precision a factor of 1,000 times better than what van de Kamp could achieve at the Sproul Observatory. This would be more than enough to find Jupiter-like planets and would enable Uranus-like planets to be detected indirectly.

About the same time that the Planetary Astronomy Committee began its deliberations, another important committee began to study the general issue of looking for extrasolar planets and the material from which they are made. In 1985, NASA requested a multiyear study by the National Research Council's Space Studies Board, who passed the job to the Committee on Planetary and Lunar Exploration (COMPLEX).

The National Research Council has numerous committees that perform detailed scientific studies for government agencies that desire independent, expert advice and are willing to pay a pittance (i.e., a scientist's ransom) for it. Unlike scientists who are paid handsomely to testify as experts in high-stakes judicial proceedings, the scientists who labor over these National Academy studies receive no financial compensation beyond their travel expenses, though they do have the dubious opportunity, during the committee meetings, to gobble up as many donuts as they can stand. Of course, they also have the chance to achieve a certain degree of immortality by seeing their names printed in the final reports. In the end, though, most scientists are suckers for the chance to dispense free advice, and this tendency often makes them objects of ridicule by engineers, who are savvy enough to know they should charge for the benefit of their advice.

No report or advice from the National Research Council committees can be made public until it is first approved by a report review committee consisting of members of the National Academy of Sciences and acting as a strict quality control inspector. As a consequence of the care taken in their production and of the scrutiny they subsequently receive by the Academy's reviewers, reports by COMPLEX and its sibling committees are generally recognized as being crucial for the success of any major scientific initiative the federal government is considering undertaking. NASA's robotic missions to the planets normally seek the blessing of COMPLEX at some point in their lives, though in the end NASA is legally, if not morally, free to do as it wishes, and at times does.

Because of the painstaking manner in which COMPLEX reports are produced, the report on extrasolar planets was not released until 1990, even though the relevant presentations were essentially concluded by 1987. The Planetary Astronomy Committee, which started its work at about the same time, was a little faster, publishing its final report in 1989. COMPLEX was chaired during most of this time by Robert Pepin, David Black's Ph.D. thesis advisor at the University of Minnesota, though Black himself was not on COMPLEX. I joined COMPLEX in mid-1990, just in time to get my name immortalized on the final report, even though all I had a chance to do was to comment on the criticisms of the report made by the Academy's anonymous quality control reviewers.

COMPLEX gave the search for extrasolar planets the necessary blessing, and then some. The COMPLEX report called for NASA's then Office of Space Science and Applications to begin a systematic search for extrasolar planets lasting a decade or more. The centerpiece of the program would be the same Space Station astrometric telescope that the Planetary Astronomy Committee had supported, capable of finding Uranus-mass planets in a nearby planetary system with the same orbital configuration as ours. In addition, COMPLEX could not resist the opportunity to ask for NASA's support for a whole shopping list of related scientific activities, such as theoretical studies on planet formation and laboratory work on primitive meteorites. The shopping list approach was the normal way in which consensus was achieved then in scientific committees composed of people with different goals and backgrounds—the committee often ended up asking for everything that anybody on the committee felt was desirable. In the go-go, boomtown years of the 1980s, when NASA's annual budget was steadily rising, that approach seemed to be appropriate.

The enthusiasm of the Planetary Astronomy Committee and COMPLEX for extrasolar planet searches was clear well before their final reports were issued in 1989 and 1990. Furthermore, NASA's Solar System Exploration Division was beginning to run out of planets that had not yet been subjected to the prying stares of robotic spacecraft on flyby orbits, largely because of the incredibly successful Pioneer and Voyager missions to the outer planets of our Solar System. After *Voyager 2* photographed Neptune in August 1989, only lonely, diminutive Pluto would remain a distant blur in a telescopic image. If NASA's planetary division did not want to succumb to the deadly "been there, done that" syndrome, it needed some new planets, and it needed them soon. By 1988, then, NASA was serious enough about searching for new solar systems that it was ready to move on to the next phase of planning. That meant it was time to form another committee.

The head of the Solar System Exploration Division in 1988 was Geoffrey Briggs, who formed the Planetary Systems Science Working

Group (PSSWG) early in that year. Briggs's letter formally appointing me as a member of the PSSWG included a rather cryptic description of the committee's charter, including to "identify and evaluate opportunities for pursuing planetary science objectives by observing stellar environments." Being out of the loop at that time on what NASA really wanted done by the PSSWG, I had no clue from the letter that the PSSWG's main task was to figure out how to find extrasolar planets. But being just another sucker for the chance to give out cheap advice and eat donuts in return, I happily joined the PSSWG, without even comprehending what it was all about.

The PSSWG was a NASA committee known as a Science Working Group. The creation of a Science Working Group means that NASA is getting serious about spending some big money, and understandably wants to make sure that it gets timely and focused advice about the expensive path it is about to undertake. The acronym for a Science Working Group (SWG) is pronounced as "swig." That tradition had the unfortunate consequence that our committee's acronym, PSSWG, soon became pronounced as the "piss-swig," to the merriment of the PSSWG members.

APRIL 11–12, 1988: The first PSSWG meeting was held in a small meeting room at the Holiday Inn Capitol in Washington, D.C., close to NASA's headquarters building. The meeting was largely devoted to deciphering what we were supposed to be doing and how we would go about doing it, but we still had time for presentations about the golden opportunity to put an astrometric telescope on the Space Station. The Space Station was being designed at the time to provide several locations along a boom where scientific instruments could be mounted as "attached payloads," and it was time to sign up and grab a spot on the boom or else miss this opportunity.

The chairman of the PSSWG was Bernard F. Burke of the Massachusetts Institute of Technology, one of the leading figures in radio astronomy and interferometry. Burke was part of a team that discovered in 1955 that Jupiter was a strong source of radio waves. The discovery was made completely by accident while Burke was a staff member at Carnegie's DTM. Burke and other DTM scientists had built a huge, X-shaped radio telescope that covered a suitably flat, 90-acre field in nearby Maryland. The radio telescope was a fixed interferometer, with a field of view that swept across the sky as the Earth rotated. Burke and the team soon noticed that about once every three days they picked up a strong burst of radio waves late at night that lasted no more than 15 minutes. At first they thought the noise was caused by some local effect, such as an automobile, until they noticed that the time at which the noise showed up was predictable and tracked along with the stars in the sky—the source of the noise could not be on the Earth. Jupiter was soon identified as the culprit.

This discovery was the first hint that planets could be strong sources of radio waves. While Burke had long since moved on to other interests, his early unintentional success in detecting planets with exotic instruments made him a natural choice for the first chairman of the PSSWG—perhaps some of his good luck would rub off on the planet detection people. Considering the dismal history of the field, they seemed to need some.

The second PSSWG meeting was held on July 12–14, 1988 at the Peppertree Inn in Santa Barbara, California—my Ph.D. advisor, Stanton Peale, had also been asked to join the PSSWG, and he volunteered the location. Burke would not willingly pass up an opportunity to hold a PSSWG meeting in a pleasant location like Santa Barbara. (Many of the locals feel it is better to be a poor person in Santa Barbara than to be a rich person somewhere else.)

At the Santa Barbara meeting, the stakes and players in the game became much clearer. Evidently, there were several different ideas for how to go about finding extrasolar planets with a specialized space telescope, and due to the obvious expense of such telescopes, it was likely that only one idea would survive, if that. The initial idea of getting a free ride and mounting a telescope on the Space Station was forgotten soon after it was realized that the Space Station would vibrate much too much to permit the delicate astrometric observations to be made—if an astronaut so much as sneezed, the Space Station would ring with microscopic vibrations that nobody would notice, except for a space astronomer trying to do precise astrometry with a telescope waving back and forth on the boom.

Most of the committee members were advocates for one space mission or another. Richard Terrile and his JPL colleagues had developed a concept for a space version of the technique that he and Bradford Smith had used to photograph the Beta Pictoris dust disk—the Circumstellar Imaging Telescope (CIT) would be a space telescope with supersmooth optics and a coronagraph to block out a star's light, allowing dust disks to be imaged with unprecedented clarity. Smith was also on the PSSWG. David Black, George Gatewood, and the University of Arizona's Eugene Levy and their team had jointly developed the idea for an

Astrometric Telescope Facility (ATF), which could fly by itself in space without being attached to the Space Station. Robert Brown of the Space Telescope Science Institute was interested in finding planets with the Space Telescope, which was almost ready to be launched into Earth orbit. Robert Reasenberg of the Harvard-Smithsonian Center for Astrophysics developed a plan for the Precision Optical Interferometer in Space (POINTS), which would find planets by the astrometric method. A competing astrometric interferometer was developed by Michael Shao of JPL, called the Orbiting Stellar Interferometer (OSI), 20 meters in length and by far the largest of the instruments proposed.

Peale and I were two of the few committee members who had no particular desire for one concept or another to win—we just wanted *somebody* to find some extrasolar planets. As theorists who had worried more about making planets than about how to find them, we were often inundated by the sudden firefights over arcane but crucial technical details of ultra-high-precision space astrometry and interferometry that soon became the hallmark of PSSWG meetings. It was clear even to us, though, that one of the best ways to advance one's own design was to point out serious flaws or limitations in a competing design. The peculiar pronounciation of the PSSWG's acronym soon became appropriate because of the battles we witnessed, with PSSWG members figuratively urinating on each other's designs for these expensive space-based instruments. While spirited, these debates were always conducted at a restrained, exceedingly rational level and did not stoop to harsh or angry words—afterward, we could all go out to dinner together and laugh about the day's interchanges and the precarious nature and prospects of hunting for planets.

8

BROWN DWARFS BY THE HANDFUL

Every great scientific truth goes through three stages. First, people say it conflicts with the Bible. Next they say it had been discovered before. Lastly they say they always believed it.

—Louis Agassiz (1807–1873)

AUGUST 8, 1988: The International Astronomical Union (IAU) was meeting in Baltimore, Maryland, holding its triennial General Assembly. Usually these far-flung meetings are devoted to reviewing past accomplishments in a novel location, but this time there was an important new discovery to announce. Harvard-Smithsonian astronomer David Latham announced that he had found a new type of evidence for a "planet" in orbit around a star with the unromantic name of HD 114762, literally the 114,762th object on a list compiled by the astronomer Henry Draper. In 1872 Draper became the first to use a photograph to capture the spectrum of a star, the same bright star (Alpha Lyrae) that *IRAS* a century later found to be so peculiar.

A star's spectrum can be examined by passing the starlight through the prism in a spectrometer, in which different colors of light are bent into different directions. A good spectrometer will show that a star's light is emitted with an intensity that depends strongly on the wavelength: The intensity will be more or less constant over some range in wavelength, and then suddenly drop (or rise) into a characteristic bell shape called a line. Lines occur because the outer layers of a star are cool enough to contain atoms that can absorb and emit light at specific wavelengths. A star's spectrum contains lines that either rise above (emission lines) or fall below (absorption lines) the general trend of radiation given off by the star (continuum emission). Latham's detection depended on the existence of these stellar lines to infer the presence of an unseen companion to HD 114762.

Latham used the *spectroscopic* method to find the companion to HD 114762. The spectroscopic method is an indirect method, like the astrometric method, that infers the presence of an unseen companion through the companion's effects on the star itself. While the astrometric method measures the position of the star as it moves about the center of mass of the star-companion system, the spectroscopic method measures the *velocity* of the star as it moves around its orbit.

The velocity of the star can be determined by using the principle of the Doppler effect, familiar to anyone who has stood close to the path of an approaching train or automobile and has heard the oncoming engine's sound suddenly drop in frequency once the engine passes by and begins receding. The sound waves given off by an approaching object are shifted to shorter wavelengths, because the object continues to move closer while it is emitting the sound waves, decreasing the distance between successive troughs and crests of the waves. A shorter wavelength is perceived by the ear as a higher frequency, so the oncoming object will sound higher pitched than it does when it is standing still. By the same type of argument, a receding object emits a sound of longer wavelength and lower frequency than the same object at rest. The amount of the shift in frequency depends on the velocity of the object.

The Doppler effect was named for Christian Johann Doppler, an Austrian physicist who pointed out in 1842 that the effect should apply equally well to light waves as to sound waves. Light coming from an object moving away from us is said to be redshifted, while objects coming toward us have their light blueshifted—red light has a longer wavelength than blue light. On the other hand, light coming from an object moving across our field of view, neither toward nor away from us, is not changed at all. Doppler showed that his effect could explain the otherwise unfathomable behavior of the light coming from binary stars.

The emission and absorption lines in the spectrum of a star provide a convenient means to measure the velocity of a star. Latham found that lines in HD 114762's spectrum would periodically be redshifted and then blueshifted, by the same amount in both directions. This shifting in frequency of the star's light was found to be repeatable, over and over again. Because the Doppler effect is only produced by motion of an object back and forth along the direction to the observer, HD 114762 had to be periodically moving first toward and then away from the Earth. This is just what would be expected for a body moving around in an orbit at more or less constant velocity, viewed from an

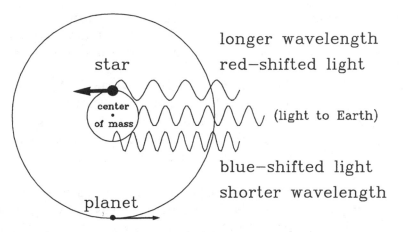

FIGURE 14. The presence of a planet can be inferred by detecting the velocity of the star around the center of mass of the system. The star's velocity along the line of sight to Earth can be measured because of the Doppler effect, in which the star's light is shifted to either longer or shorter wavelengths by an amount proportional to the star's velocity. (Courtesy of Alan Boss.)

angle close to being in the same plane as the star's orbit—at one end of its orbit, the star moves toward the Earth, while at the other end it moves away from Earth. The amount by which the star's lines are red-shifted or blueshifted depends on the masses of the star and the companion, the period of the orbit, and the viewing angle.

Latham knew that HD 114762 was a star that was similar to the Sun in mass, and from examining his data he could tell that the period of the orbit was 84 days, about the same as Mercury's orbital period around the Sun. Given the magnitude of the velocity variations that he found, from 600 meters/second toward Earth to 600 meters/second away from Earth, Latham could infer that HD 114762 was orbited by an object with a mass perhaps as low as 11 Jupiter masses, well into the suspected brown dwarf regime.

The true mass of the companion could be known only if the angle between the plane of the companion's orbit and the direction to the Earth was known. The lower limit of 11 Jupiter masses would be the correct value if the direction to the Earth happened to be aligned with the plane in which HD 114762 and its companion orbited, so that the entire Doppler shift was radiated in the Earth's direction. A Doppler shift is only produced to the extent that the motion of the source is toward or away from the observer, and this is maximized when the observer lies in the orbital plane. If the Earth was close to looking straight down on the star-companion's orbital plane, on the other hand, the mass of the companion could be as much as that of a normal star, because then most of the motion would be across the field of view and would produce no Doppler shift.

Latham began monitoring HD 114762 in 1981 not because he was looking for planets, but because he wanted a "standard star" with no velocity variations that could be used to prove that his system of searching for binary stars would not introduce spurious signals when nothing should be found. By chance, HD 114762 turned out to be a nonstandard star indeed.

Latham's spectrometer produced errors on the order of 400 meters/second, not much smaller than the signal he was detecting. Hence Latham sought corroboration of his detection by another astronomer, using a completely different telescope and spectrometer—Michel Mayor of the Geneva Observatory in Switzerland. Mayor had a spectrometer with a precision comparable to that of Latham, and Mayor began to take intensive measurements of HD 114762 soon after Latham alerted him in 1988. Mayor's observations fully confirmed Latham's solution, and between the two of them they observed HD 114762 enough times to be sure that they had a sound detection.

By the time of the August IAU meeting, Latham was ready to go public. Latham pointed out that while the viewing angle was uncertain, the fact that HD 114762 could have a companion as small as 11 Jupiter masses meant that it could have a planetary companion—after all, I had presented results at the 1985 George Mason conference saying that the minimum mass of a brown dwarf star was in the range of 20 to 50 Jupiter masses. Latham's minimum mass fell comfortably below this theoretical prediction, so if the companion was too small to be a brown dwarf, then perhaps it had to be a planet. *Science* magazine reported in that week's issue that a "planet" had been found around HD 114762 and quoted Latham as saying that "this is the first good orbital solution for a single body the size of a giant planet." The fact that the "planet" orbited so close to its star (at about 40 percent of the Earth-Sun distance, at Mercury's orbit) was noted as a potentially troubling aspect—giant planets were not expected to be found so close to their stars, based on the analogy of our solar system.

By the time (early 1989) that Latham and Mayor and their colleagues were ready to submit their results to *Nature* for publication, however, they cautiously decided to phrase their discovery as a probable brown dwarf rather than as a planet, and noted that because of the unknown viewing angle, the companion's mass could well exceed 11 Jupiter masses.

Their caution in not claiming the detection of a planet was well founded, because they evidently were not aware that I had revised downward my 1985 estimate of the minimum mass of a protostar. In a paper published in the *Astrophysical Journal* in 1986, I gave a value of about 10 Jupiter masses as this minimum mass, based on a more complete set of theoretical models of cloud collapse and fragmentation than I had been able to complete by the time of the George Mason conference. Subsequent papers in the series revised this minimum mass both upward and downward, with the net effect being that in the end, my best estimate of the minimum mass of a brown dwarf star was about 3 Jupiter masses. Latham and Mayor's object was clearly well above this limit, and so HD 114762 looked to be the first bona fide brown dwarf star.

Discoveries of more brown dwarf stars soon began to leap out of the woodwork.

JUNE 14, 1989:   Rochester's William Forrest, who had earlier helped to show that the brown dwarf VB8 B did not exist, announced the discovery of *nine* brown dwarfs of his own during a talk at the American Astronomical Society meeting in Ann Arbor, Michigan.

Forrest and his team used his infrared camera at the Hawaiian IRTF to photograph regions in the Taurus molecular cloud. Young stars were known to lace the Taurus region, and so Taurus was a likely place to find young brown dwarf stars; by virtue of their youth they would be brighter and easier to find than older brown dwarfs. By looking around 27 known T Tauri stars, Forrest and his team expected to find faint, red objects that could be brown dwarfs—and they found a bunch. Some of the objects could even be shown to be moving in the same direction on the sky as the T Tauri stars, implying that these dim objects might truly be located in the Taurus cloud along with the T Tauri stars. If these faint objects were not in Taurus, they might well be much more distant, normal stars that only appeared to be faint and red because they were so far away and their starlight had been reddened by passage through intervening clouds of interstellar dust.

The announcement was hailed in the *Washington Post* the next day as being "the real thing," in comparison to the unforeseen dissolution of the VB8 B discovery. Assuming an age of a million years, Forrest and his team found that the objects would have to have masses in the range of 5 to 20 Jupiter masses in order to be that faint at a young age.

Science confirmed that the exasperating quest for brown dwarf stars may have finally ended with Forrest's findings but pointed out that there were dissenting opinions. Benjamin Zuckerman stated that the objects were likely to be low-mass stars, not brown dwarfs, because the objects gave off too much light at visual wavelengths to be brown dwarfs. Zuckerman had vented some of his opinions of Forrest's results during a review talk at the Centennial Meeting of the Astronomical Society of the Pacific, held in the lecture halls of the University of California, Berkeley the week after the Ann Arbor meeting and had even circulated copies of his personal letter to Forrest to 18 other scientists the next week, laying out his problems with Forrest's interpretation. Zuckerman felt that Forrest had found nine background stars, not nine brown dwarfs.

Whether Forrest and his group or Zuckerman were right could only be determined by further observations. The key would be to search the spectrum of Forrest's brown dwarfs for the telltale emission and absorption lines that would reveal whether the objects really were cool brown dwarf stars or much hotter normal stars.

AUGUST 26, 1989: The existence of a *binary* brown dwarf system was proclaimed by Swarthmore's Wulff Heintz, working with the same Sproul Observatory telescope employed in van de Kamp's lifelong efforts. Heintz was well aware of the astrometric telescope problems that led to the downfall of the planets around Barnard's star, and he made doubly sure that his measurements would not be subject to the same problems that slowly crept up on the Barnard's star work. In fact, Heintz was able to show that part of the problem with van de Kamp's data lay in the photographic plates themselves, and not just with the telescopes. Heintz showed that different batches of photographic plates, being subtly different in the way they were manufactured, would respond in different ways to the same stellar image, forming slightly noncircular shapes that shifted the apparent center of the image (and hence the star's apparent position) for no good reason at all.

Wary of these severe problems and determined to avoid them, Heintz submitted a one-page paper to the European journal *Astronomy and Astrophysics* with his analysis of 50 years of data for the Wolf 424 star system. Wolf 424 was a pair of stars just far enough apart that their combined image looked suspiciously distorted, like two basketballs sewed inside a burlap sack. Heintz had been following the motions of this binary system for many years, watching the balls move around each other inside the burlap sack, and by 1989 he was sure that he had a good grip on the orbit and could estimate the masses. As Heintz expected, based on his previous work, Wolf 424 seemed to have a period of 16 years, which implied that the total mass of the system was only 110 Jupiter masses. Heintz decided that Wolf 424 contained two brown dwarf stars with masses of 50 and 60 Jupiter masses apiece. Heintz had found *two* brown dwarfs for the price of one. Coincidentally, I had published a letter in *Science* two years before saying that the best place to find a brown dwarf star was in orbit with another brown dwarf, and Heintz seemed to have accomplished this feat.

But other astronomers were not finding brown dwarfs at all. At the same Baltimore IAU meeting where Latham presented the case for HD 114762, Bruce Campbell of Canada's Dominion Astrophysical Observatory had argued that solar-type stars generally do not have *any* companions in the range of 10 to 100 Jupiter masses—brown dwarfs simply could not be found in the small sample of 14 stars he had studied. Campbell and Gordon Walker of the University of British Columbia were two of the pioneers of applying high-precision spectroscopy to search for low-mass companions; in the late 1970s they had developed a technique that used the corrosive and deadly gas hydrogen fluoride to serve as a precise spectral benchmark against which to search for tiny, periodic Doppler shifts indicative of the presence of companions. Their measurements were over a factor of 10 times more precise than those that Latham and Mayor used to find HD 114762 and so should have uncovered any brown dwarfs lurking in orbits close to the stars in their sample.

In fact, two years before, Campbell and Walker had claimed at an American Astronomical Society meeting that while they had not found any brown dwarfs, they did see signs of a slight periodic Doppler shift in two of their stars, Epsilon Eridani and Gamma Cephei. The detections were marginal but predictably were greeted with enthusiasm by the press, with headlines again proclaiming the discovery of extrasolar planets. By the time that Campbell and Walker were ready to submit their results for publication in the *Astrophysical Journal,* however, they no longer firmly believed in a planetary companion for Epsilon Eridani (van de Kamp's earlier claim for a planet around Epsilon Eridani had since been disproven by Heintz). Campbell and his colleagues still maintained, though, that Gamma Cephei might have a planet with a mass no smaller than 1.7 Jupiter masses.

Geoffrey Marcy of San Francisco State University also reported finding no evidence for brown dwarf stars in 1989. Marcy had started searching for the minute Doppler shifts of stellar wobbles in 1983 when he was a postdoctoral fellow at the Carnegie Institution's Mount Wilson Observatory. He continued the program as a visiting astronomer at

the University of California's Lick Observatory, moving up an incre-
ment from Mount Wilson's 100-inch-diameter telescope to Lick's 120-
inch reflector. While his precision was not as good as that of Campbell
and Walker, Marcy could still hope to find brown dwarf stars, if they
were common around the 65 low-mass stars he observed. Because of
the low masses of the primary stars (typically just a third that of the
Sun), very-low-mass companions should have been even easier to see.

Marcy found nothing particular to speak of, at least nothing with
an orbital period shorter than the four years for which he had been taking data. Late in 1989 Marcy published his disappointing results in the
Astrophysical Journal, to the chagrin of theorists like the University of
Arizona's Jonathan Lunine, who with his colleagues had been working
for years on calculating the likely appearance of brown dwarfs, objects
that might not even exist if Campbell's and Marcy's negative results
were representative of most stars.

JANUARY 1–3, 1990: John Stauffer of the Harvard-Smithsonian Center
for Astrophysics and his team used a spectrograph on the 200-inch
Hale Telescope to measure the optical spectra of six of Forrest's brown
dwarf candidates. The results were disastrous for the original interpretation. None of the six objects showed evidence of absorption due to
molecules such as water that could only exist in the cool outer layers of
a brown dwarf star. The objects appeared to be mundane solar-type
stars lying far behind the Taurus molecular cloud, and heavily reddened
as a result. Forrest was a member of Stauffer's team, and he agreed with
the sad conclusion that the new spectra had suddenly erased most of his
nine brown dwarfs from the slate. Zuckerman's pessimism turned out
to be correct.

9

A FRESH START

Never say, "I tried it once and it did not work."

—Lord Ernest Rutherford (1871–1937)

APRIL 24–29, 1990: NASA's Space Telescope was finally launched into low Earth orbit by the space shuttle *Discovery*, after having been delayed several years by the disastrous explosion and loss of the space shuttle *Challenger* in 1986. Named for Carnegie Institution astronomer and cosmologist Edwin P. Hubble, discoverer of the expanding universe, the *Hubble Space Telescope (HST)* carried a 94-inch primary mirror that was expected to produce astronomical images of unprecedented clarity, free of the distortions of the Earth's atmosphere. With several billion dollars invested in building *HST,* NASA was relieved to see *HST* safely launched and placed in the proper orbit, ready to be checked out and then to begin science operations.

No one was more relieved than Edward J. Weiler, NASA's Chief Scientist for *HST.* Weiler had been shepherding the project for years, and he knew well that *HST* would produce spectacular results if given the chance. In fact, *HST* was expected to be the first major improvement in telescopic power since the completion of the 200-inch Hale Telescope in 1948.

While *HST* was designed with general astronomical observations in mind, there was also hope that it would find extrasolar planets, but not through the lunar-orbit maneuvers proposed by the Project Orion team. High-precision astrometry would be possible with *HST,* to an accuracy (roughly 1 milliarcsecond) about a factor of 10 times better than was possible at van de Kamp's Sproul Observatory. The *HST* Astrometry Team at the University of Texas in Austin estimated in 1982 that about a dozen nearby stars could be profitably searched for companions by *HST.* In anticipation of *HST's* possible role in discovering new planets, the Space Telescope Science Institute in Baltimore held a major conference in 1988 devoted to the subject of planet formation and detection.

However, by the time that *HST* was finally launched in 1990, George Gatewood had developed a new technique for performing high-precision astrometric measurements from the ground and was able to achieve the same 1-milliarcsecond accuracy as *HST* using the venerable 30-inch Thaw refractor of the Allegheny Observatory, outside Pittsburgh, Pennsylvania. Gatewood had done away with the photographic plates that had helped to deceive van de Kamp and had built a clever device that collected the star's light with modern electronic detectors and measured the star's position with respect to its neighbors with a finely ruled reference grating. Gatewood's device would be limited only by the turbulence of the Earth's atmosphere, not by uncontrolled errors on the ground. The fact that *HST* could do astrometry no better than what could be done from the ground in western Pennsylvania underlined *HST's* limited usefulness for finding extrasolar planets, a purpose for which it was not designed. In fact, the *HST* astrometry was to be performed with fine guidance sensors, whose primary reason for existing was to control the pointing of the telescope, not to do ultrahigh precision astrometry.

The elation of Weiler and others at NASA Headquarters over the successful launch of *HST* turned to deep depression a few months later, when it was decisively shown that the primary mirror on *HST* had been manufactured incorrectly. A decade earlier, a technician had inserted the wrong end of a measuring rod into the device that measured the shape of the *HST* mirror as it was being ground out of a flat cylinder of glass—as a result, the mirror was ground into a shape that focused light not at a single focal point, but along a line 3.8 centimeters long. Very blurry images resulted. Because of this severe spherical aberration, which went undetected until after *HST* was in orbit, *HST* was unable to begin taking the extraordinarily sharp images that were its main reason for being. Looking for planets with *HST* suddenly went down the tubes. Any planet searches would have to wait for *HST's* vision to be corrected by a specially designed optics system, to be installed three years later in a daring space walk by astronauts from NASA's space shuttle *Endeavour,* after which *HST* would finally be able to send to Earth the incredible and glorious astronomical images for which it was designed.

A few months later in 1990, two Space Telescope Science Institute astronomers published a thorough examination of *HST's* chances for detecting extrasolar planets through direct imaging of the regions around nearby stars. Robert A. Brown and Christopher Burrows showed that there was no hope for extrasolar planets to be detected with *HST,* even after the spherical aberration of the primary mirror was fixed. Light from the star would still be scattered into the location of the image where the planet would appear, swamping the planet's much fainter light. *HST* simply was not designed with the excrutiatingly difficult problem of detecting planets in mind, and it could not be coaxed into performing the miracle. Something else would be needed.

AUGUST 27, 1990: The PSSWG had been charged by NASA Headquarters with developing a scientific strategy to search for and find new solar systems. A subset of the PSSWG presented the committee's preliminary plan to Lennard Fisk, then NASA's Associate Administrator for Space Science and Applications, in a meeting at NASA Headquarters. Fisk's support was crucial—he was the one with the power to sign letters and to commit NASA to expensive programs. There was a very expensive program that we wanted Fisk to approve.

The PSSWG had put together a three-phase plan to find new planets, whimsically named the TOPS plan by the chair, Burke, with the acronym standing for Toward Other Planetary Systems. Most of the TOPS plan had been formulated during a week-long workshop in January 1990 at the old Lunar and Planetary Institute in Houston, whose new director was PSSWG member David Black. However, the driving element of the first phase of the TOPS plan had arisen unexpectedly out of the dark blue waters of the Big Island of Hawaii after the Houston workshop.

The William Keck Foundation had stunned the philanthropic world by giving over $70 million to the University of California and Caltech to build the world's largest telescope on the summit of Mauna Kea. The gift was by far the largest single donation on record. But by early 1990, backed by further promises of money from the Keck Foundation, the ambitious UC/Caltech partnership was ready to consider building a second mammoth telescope—there would be not one, but *two* Keck telescopes. Each Keck telescope would be 10 meters in diameter, and the two would be located about 85 meters apart in a single complex, which would enable them to be combined into an optical interferometer with incredible light-gathering power. At an elevation of almost 14,000 feet in the middle of the normally placid Pacific Ocean, the Keck Observatory would enjoy one of the best sites for astronomical observation on the planet.

To build the second Keck telescope, however, UC/Caltech needed a partner to come up with about $35 million, about one third of the cost of the second telescope, Keck II. In return, the partner would receive one third of the observing time on Keck II or, equivalently, one sixth of the time available on either Keck I or Keck II.

Edward Stone, the director of JPL and head of the UC/Caltech consortium building the Keck Observatory, offered NASA the chance to buy into this limited-time, no-money-down opportunity. NASA had two different divisions that might be interested, one that focused on traditional astronomy and the other on the Solar System, groups that historically had been rivals at NASA and elsewhere. NASA's Astrophysics Division, leery of breaking the pact with the NSF and getting involved with a major ground-based observatory, declined Stone's offer. However, the planetary scientists supported by NASA's Solar System Exploration Division, already used to using the IRTF on Mauna Kea to observe Solar System objects, leapt at the chance to become significant users of the Keck Observatory. The Keck Observatory might or might not turn out to be useful for finding new planets, but it would certainly be dandy for studying the inhabitants of our Solar System.

Because the PSSWG's search for new solar systems was optimistically expected to lead to new pots of money for NASA, by virtue of the intrinsic interest of finding new worlds, it fell to the PSSWG to justify NASA's involvement with the Keck Observatory. That way NASA's share of the Kecks would not have to be paid for by taking money from existing NASA programs, which would surely create a deadly firestorm of resistance from those whose programs were perceived to be threatened. If *anything* was going to induce the Administration and the Congress to augment NASA's space science funds, surely the race to find extrasolar planets was it.

We presented the TOPS plan to Fisk, which involved the Keck Observatory as the first major step along the road to a serious planet detection effort by NASA. This first step would also include expanded support for existing ground-based efforts, such as those of Gatewood and Marcy. The second step would be a space-based telescope along the lines of the various competing proposals that the PSSWG had already been battling over. The third step was poorly defined at the time, but might consist of something truly visionary, such as a huge interferometer on the backside of the Moon, operated by NASA astronauts.

The Keck Observatory people needed a quick answer from NASA if they were to keep their project on track. Fisk listened to our presentations and accepted the PSSWG's arguments; soon thereafter Fisk signed the letter to Stone committing NASA to a partnership in the Keck Observatory—phase 1 was now underway. The PSSWG was generally pleased to see their proposed TOPS program off to a running start, due in large part to the unexpected Keck opportunity, even if the Keck telescope was not most people's idea of the ideal planet finder.

DECEMBER 4, 1990: After 13 years of labor, two seasoned Swiss astronomers submitted the results of their exhaustive search for binary star companions to *every* star similar to the Sun that they could find within 72 light years of the Earth. Michel Mayor, codiscoverer with David Latham of the possible brown dwarf companion to HD 114762, and Mayor's Geneva Observatory colleague Antoine Duquennoy had used every type of binary star data available, in addition to making many spectroscopic observations of their own to look for unseen companions. Duquennoy and Mayor's survey definitively showed that single stars like the Sun are relatively rare—two out of three solar-type stars have at least one other stellar companion.

Even more interesting, Duquennoy and Mayor's survey conclusively showed that the orbits of binary stars had a key identifying trait. With one exception, the binary stars in their sample were always found to have *eccentric* (elliptical) orbits that were not even close to being circular, a powerful clue to their origin.

The exceptions to Duquennoy and Mayor's rule were the binary stars that had short orbital periods, circling each other in less than 11 days. Duquennoy and Mayor found that these systems had nearly circular orbits. However, their short orbital periods meant that these stars were also close to each other, close enough for their gravity to exert significant tidal forces on each other, just as the Moon exerts on the Earth. As a result, the close binary systems were expected to interact tidally in such a way that if their orbits had ever been eccentric, tidal forces would eventually turn their orbits into circles. The closer together the stars were, the faster this process of tidal circularization would be. During the several-billion-year lifetime of the average solar-type star in the Sun's neighborhood, there would be enough time to circularize the orbits of stars as far apart as those with periods of about 11 days. Binaries with even longer periods (and greater separations) would be largely unaffected by the much weaker tidal forces, which depend strongly on distance. Binaries with periods much greater than 11 days would then retain the characteristics of the orbits on which they originally formed.

Duquennoy and Mayor had thus presented strong evidence that *all* binary stars had eccentric orbits, unless they happened to orbit so close to each other that tidal forces had time to circularize their orbit. Wide binary stars stay more or less on the same orbit forever, so if they were eccentric now, they must have been formed on an eccentric orbit billions of years ago. Any scheme for explaining the formation of binary stars must therefore explain why they began their lives so eccentrically.

This was a dramatically different situation from that of the major planets of our Solar System, in which Earth, Venus, Jupiter, Saturn, Uranus, and Neptune all have orbits around the Sun that are reasonably close to being circles. Because gravitational forces between the planets over the 4.6 billion years since they were formed could only serve to increase their orbital eccentricities, it was clear that the major planets of our Solar System must have formed on nearly circular orbits if they were still on circular orbits at the present.

Theorists believed that they understood the reasons for the nearly circular orbits of the planets. One of the triumphs of the detailed models of terrestrial planet formation computed by George Wetherill was that Earth and Venus, even though assembled out of collisions with a swarm of smaller bodies buzzing around on noncircular orbits, inevitably ended up themselves on nearly circular orbits. This was because Earth and Venus had been knocked in so many different directions, so many times, that they woozily ended up walking a more or less straight line, which in their case really meant a circular orbit. The fact that the much smaller terrestrial planets, Mercury and Mars, have somewhat more eccentric orbits could be easily explained by their not having digested enough other planetesimals to have averaged out to existence on a circular orbit.

As for the giant planets, Peter Goldreich and Scott Tremaine's analysis of the interactions of giant planets with the disk gas from which they gained the bulk of their mass had shown that these interactions should rapidly circularize the orbits of the planets, should they ever have been noncircular. Hence there was good reason to believe that the major planets of our Solar System formed on roughly circular orbits and would remain on roughly circular orbits. By analogy, the same was to be expected for planets that formed around other stars.

Because both indirect methods of detecting unseen companions yield estimates of the orbital eccentricity, there was thus a real potential for using orbital eccentricity to judge whatever had been found—was it really a planet, or was it a brown dwarf star? Planets would have circular orbits, whereas brown dwarfs, being severely mass challenged but stars nonetheless, would have eccentric orbits.

Duquennoy and Mayor's survey also yielded tantalizing evidence for a reasonably large population of brown dwarf companions to solar-type stars, a result that was in direct variance with the earlier null results of Campbell and Marcy, albeit for smaller samples of stars.

Duquennoy and Mayor found evidence that about 8 percent of the Sun-like stars in our neighborhood of the galaxy were orbited by brown dwarfs or low-mass stars with masses in the range of 10 to 100 Jupiter masses. They also found that the suspected companions appeared to have eccentric orbits, just like heftier binary star companions. The observational rule that eccentric orbits imply a stellar origin could thus be firmly extended to include likely brown dwarfs.

Interestingly, both Latham's suspected brown dwarf companion to the normal star HD 114762 and the brown dwarf binary studied by Heintz, Wolf 424, had eccentric orbits, nicely consistent with the orbital eccentricity acid test advanced by Duquennoy and Mayor.

Duquennoy and Mayor concluded that their survey thus supported the idea that binary stars, whether composed of normal stars or brown dwarfs, were formed on eccentric orbits, probably through the process of fragmentation. I was elated by this conclusion.

Peter Bodenheimer and I had initiated computational studies of the fragmentation process over a decade earlier, and I had continued working steadily on fragmentation in the interim, so fragmentation was a mechanism near and dear to my heart. I knew my calculations of the collapse of interstellar clouds and their subsequent fragmentation into protostars would be useful not only for estimating the minimum mass of a star, but also for predicting the shape of the orbit on which a protostar forms.

A few months before the Swiss survey was submitted for publication, I had analyzed the orbital properties of a number of protostars formed by the fragmentation process. The results were as I had expected—the protostars in my models nearly all began their lives on eccentric orbits, consistent with Duquennoy and Mayor's survey. The result was expected because the fragmentation process occurs during the sudden collapse of a cloud, when there is no way to achieve the delicate balance of gravitational and centrifugal forces that is needed to produce a circular orbit. Instead, the newly formed protostars tended to fall in toward each other and then fly back out again, awkwardly trying to find a stable arrangement, like a newlywed couple with sudden misgivings.

Ian Bonnell and his colleagues at the University of Montreal reached the same conclusion. Bonnell's models were calculated with a numerical technique that treated an interstellar cloud as a collection of particles, a technique different from my own, which represented the cloud as a fluid. The fact that we independently arrived at the same conclusion, using two different approaches, gave us confidence that we had gotten the right answer.

While Bonnell and I agreed that brown dwarfs were likely to form on eccentric orbits, there was yet another possibility for explaining

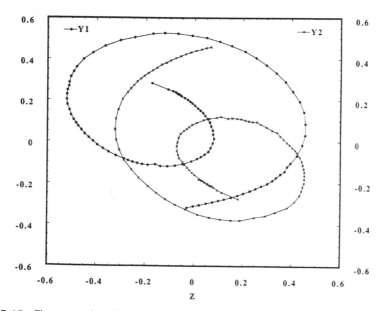

FIGURE 15. The eccentric orbits of two newly formed protostars (Y1 and Y2) formed by fragmentation out of a collapsing cloud of interstellar gas and dust. Binary stars form on noncircular orbits. (Reprinted, by permission, from I. Bonnell et al., 1991, *Astrophysical Journal,* volume 377, page 555. Copyright 1991 by the American Astronomical Society.)

Duquennoy and Mayor's survey results. Pavel Artymowicz and Steven Lubow of the Space Telescope Science Institute and their colleagues had been studying what happens to a binary orbiting inside a flat disk of gas. Using a computer code similar to that of Bonnell's, they showed that gravitational forces between the stars and the surrounding disk could lead to spiral arms in the disk that would give a kickback to the stars, making their orbit even more eccentric. Artymowicz and Lubow's model showed that even if binary stars should somehow avoid being formed with the eccentric orbits predicted by Bonnell's and my fragmentation calculations, the disk left over from their formation should make them eccentric anyway. Either way, then, binary star orbits were going to be eccentric. Theory and observations agreed.

FEBRUARY 14, 1991: The formation of solar-type binary stars was being clarified, but the water in the brown dwarf pool was getting muddy. William Cochran and his colleagues at the University of Texas at Austin submitted a letter to the *Astrophysical Journal* that threatened the status of the companion to HD 114762 as a brown dwarf star.

Cochran had started a program of searching for extrasolar planets in 1987 at the University's remote McDonald Observatory, located in west Texas. Cochran decided to use the spectroscopic method, employing the existing spectrograph on the Observatory's 2.7-meter-diameter telescope. To prevent any accidents, Cochran avoided the dangerous hydrogen fluoride gas used by Campbell and Walker to provide a reference wavelength for their Doppler shift measurements. Cochran and his colleague Artie Hatzes started out using oxygen in the Earth's atmosphere as a reference, and they found that seasonal changes in the west Texas wind limited their accuracy to about 12 meters per second—not great, but still far better than the precision that Latham had used to find the companion to HD 114762. To eliminate this atmospheric noise, in October 1990 Cochran and Hatzes switched to using a cell of gas placed right in front of the entrance to their spectrograph, as had Campbell and Walker. Unlike Campbell and Walker, however, Cochran and Hatzes used iodine gas, a much more friendly and benign substance than hydrogen fluoride. With this set-up, they could measure Doppler shifts with a precision of about 7 meters per second. Geoff Marcy and his postdoctoral fellow, R. Paul Butler, had begun using an iodine cell at the Lick Observatory in the spring of 1990—iodine was becoming the wavelength reference of choice.

Cochran and his team began to monitor HD 114762 in 1988, soon after Latham announced his startling discovery. To avoid having to pronounce HD 114762's tongue-twister of a name, Cochran and his colleagues gave HD 114762 a much easier to say nickname—George. After gathering high-precision data on George for two years, Cochran and his group submitted their results. They agreed with Latham and Mayor that HD 114762 had a companion on an eccentric orbit, consistent with the companion being a star of some sort, thus confirming that Latham and Mayor had indeed found something real.

However, Cochran and his colleagues also argued that the star HD 114762 was rotating in an unusual manner, with either its north or south pole pointed almost directly at the Earth. They deduced this orientation from their measurements of the shape of a line in the spectrum of HD 114762, which indicated that either the star was being observed nearly pole-on (a line of sight along the star's rotation axis), or else the star was rotating hardly at all. Given that other stars similar to HD 114762 were known to be rotating, just as the Sun does, it then seemed that HD 114762 must be lined up with its rotation axis pointing straight at the Earth. The Sun's rotation axis is nearly perpendicular to the plane in which the Solar System's planets orbit, so if the Solar System was any guide, then HD 114762's companion might be orbiting in a plane that was nearly perpendicular to the line of sight to the Earth. Because of the

peculiarities of the Doppler technique, this would mean that the companion had to be much more massive than the bare minimum of 11 Jupiter
masses. HD 114762 had a companion all right, but Cochran's argument
implied that it could just be a normal star, not a brown dwarf.

There was a good chance that George was an imposter. HD 114762
remained on the list of possible brown dwarf stars, along with the Wolf
424 binary, but the number of generally accepted brown dwarfs once
again shrank to zero.

JUNE 19, 1991:   We learned at a Washington meeting of the PSSWG
that the competition for the second phase of the TOPS plan, the orbiting
planet-search telescope, had become much more intense than we hoped.

The news came from Wesley Huntress, an amiable, soft-spoken,
former JPL scientist who had replaced Geoffrey Briggs as director of
the Solar System Exploration Division at NASA the year before.
Huntress told the PSSWG that there was a second NASA advisory committee that was pushing for an orbiting telescope, called AIM (Astrometric Interferometry Mission). AIM's main purpose was to use the astrometric technique not to search for planets, but to perform key
measurements of great interest for astronomy as a whole, such as measuring the distance to certain stars (Cepheid variables) that are used as
measuring rods for the size of the universe. AIM would be presented as
a top choice for a new NASA space mission by NASA's Astrophysics
Division in the upcoming Woods Hole Shootout, at which NASA's
managers and scientific advisors would gather in splendid isolation on
Cape Cod to pick the winners and losers for the next five years.

AIM would be in direct conflict at Woods Hole that summer with anything the planetary division proposed, and the fact that AIM's capabilities
(if not its goals) were similar to those of the TOPS program's orbiting telescope meant that it was clear that both concepts could not survive—there
could be at best one survivor, if even that, when the time came to choose.

The high-flying days of the 1980s were over, and the huge federal
budget deficits run up by the Reagan Administration were beginning to
frighten Congress into threatening to cut even NASA's budget severely.
Long-approved missions of great scientific interest and merit, such as
Comet Rendevouz/Asteroid Flyby (CRAF), the first mission planned to
visit the two most primitive types of bodies in the Solar System, were in
imminent danger of being canceled after years of planning and development. A bruising public battle over the fate of NASA's Space Station had
just gone badly for the scientists who spoke out against the Station; the
supporters of the Station won, and there was fear that revenge would

surely follow. NASA's administrator, Rear Admiral Richard H. Truly, was a former Navy pilot, NASA astronaut, and director of the Space Shuttle program and could not be expected to be overly sympathetic to those misguided souls who for some reason preferred robots in space to human beings. In this desperate environment, the chances of starting any new NASA science programs at all seemed to be distinctly poor.

We were assured that morning, however, that even though CRAF might be in mortal danger, the TOPS program was right on track—the money for the Keck II telescope would be there when it was needed. I skipped the afternoon session of the PSSWG meeting, because I had to return to DTM for an even more important event. George Wetherill was stepping down as director of DTM after 16 years, and a picnic was held in his honor that afternoon on the grounds of DTM. The rest of the PSSWG members spent much of the afternoon discussing strategies for the Woods Hole Shootout later that summer.

Unlike the TOPS program, which was still being incubated, the AIM proposal had received key support earlier in 1991 when a committee of the National Research Council released the results of its once-in-a-decade ranking of Big Ticket items for U.S. astronomy. Following tradition, the report was informally named after the committee's chair, John Bahcall of Princeton's Institute for Advanced Study, a theoretical astrophysicist with wide-ranging interests. Bahcall had had the honor of serving as the summary speaker at the somewhat premature workshop on brown dwarfs held at George Mason University in 1985, when VB8 B still seemed real.

The Bahcall report included AIM as one of its prioritized projects for the next decade and called it out for special note in the text of the Executive Summary of the report. However, AIM was ranked fairly far down the list, with many expensive projects ranked higher. The TOPS program did not exist when the Bahcall committee was deliberating and so did not appear at all in the report. The Bahcall report was expected to set the agenda for U.S. astronomy for the 1990s, and the fact that AIM was in the report, but TOPS was not, did not bode well for TOPS.

AUGUST 26–31, 1991: Victor Safronov and the scientists of the Schmidt Institute of Physics of the Earth planned an international conference in Moscow to commemorate the 100th anniversary of Otto Schmidt's birth.

One week before the conference was to begin, bloody street battles erupted in Moscow, and the August Revolution was underway. Safronov began the process of formally canceling the conference, but before the process could be completed, the three-day-long coup against Mikhail Gorbachev failed, and calm returned to the streets of Moscow.

A dozen adventurous American, German, and Japanese planet makers attended the conference and found that Moscow had been transformed into a city of disfigured Soviet statuary.

The Soviet Union was in the process of turning back into Russia. Most scientific efforts, even those of great achievement like Safronov's old research group, would soon find it almost impossible to survive in the new Russia. The conference effectively marked the end of the pioneering Schmidt-Safronov school of planet formation.

SEPTEMBER 6, 1991:   The critical Woods Hole Shootout had occurred, and it was time to announce the winners and losers. The Shootout had lived up to its name, with highly agitated advocates debating the merits of their favorite missions in a verbal free-for-all. The stakes were as high as they get in NASA space science—Woods Hole was no place to be modest and shy.

The Shootout turned out to be a measured success for the TOPS program. In a congratulatory letter to the PSSWG membership, Burke noted that the first phase of the TOPS program had been assigned the highest possible priority for a new start in the area of enhancing research and technology. The assignment to that category meant that the first phase of the TOPS program was viewed at Woods Hole primarily as a means to provide access (at NASA's discretion) to the otherwise private Keck Observatory.

The second phase of TOPS, the orbiting telescope that was in direct competition with AIM, did not do well at all, but then neither did AIM. Both were relegated to the category of also-rans, a jumble of unranked missions that would have to wait many years before they could be considered again. Still, at least the first phase of TOPS was properly anointed at Woods Hole, and Burke declared victory on that front.

OCTOBER 28–29, 1991:   In deference to the chair's choice of acronyms, the PSSWG had changed its unfortunate acronym to the much better sounding TOPSSWG. The TOPSSWG met at JPL and discussed how to go about "downselecting" (i.e., terminating) the various concepts still being studied for the orbiting telescope.

Reasenberg's POINTS and Shao's OSI were competing against each other to become the Bahcall report's AIM at the same time that we were discussing their abilities to find planets. It seemed pointless to continue to spend precious money developing both ideas, so a downselect, however painful, was in order.

Reasenberg and Shao were both extremely able technologists and debaters. In fact, Reasenberg and Shao used to be close colleagues at the Smithsonian Astrophysical Observatory before Shao moved on to JPL. Back at the time of the Ames meetings in 1976, Shao, Reasenberg, and Gatewood realized that it was highly unlikely that they would each achieve their separate dreams of building an orbiting telescope to find planets. The trio made an informal pact to keep each other involved in the planet search game, even if only one of them should hit the jackpot and succeed in getting the money needed to build such an expensive instrument. This shared history helped to reduce the unproductive friction that might otherwise have developed between the three.

The TOPSSWG spent several hours touring various laboratories at JPL, including Shao's OSI development lab, and heard presentations from JPL people about the plans already underway and decisions being made at JPL. It became clear that JPL, threatened with gradual downsizing by the deleterious effect of the federal budget deficit on NASA, was ravenously intending to swallow the TOPS program to help keep its army of engineers fed. It was beginning to look like there would be little room for involvement by scientists from outside JPL. Given that the TOPSSWG was full of non-JPL scientists who hoped to become a part of the TOPS effort, this realization led to a special closed-door session of the TOPSSWG, at which harsh comments were made about JPL's assumption that it would be running the TOPS program. In the euphemistic language of Washington, a "frank exchange of views" was held during the tumultuous executive session. As a result, the TOPSSWG politely and diplomatically asked NASA's Huntress to ensure that non-JPL scientists would be an "integral component" of the TOPS program.

The JPL meeting also produced the first reports that the summit of Mauna Kea might be an especially good location for astrometric planet searches. Studies by Shao and others were beginning to show that the atmospheric turbulence over Mauna Kea was even better behaved than had been expected, possibly permitting a major improvement in the ability to find planets without the huge cost of going to space. NASA's involvement with the Keck Observatory was beginning to look like a sound decision. The TOPSSWG needed to schedule a lengthy meeting in a secluded location in order to finish writing the report describing the TOPS program, and Burke took advantage of the Keck connection to propose meeting in Hawaii, close enough to Mauna Kea for the TOPSSWG to tour the Keck Observatory then under construction. The TOPSSWG enthusiastically concurred; we weren't getting paid for our years of work on the TOPSSWG, but at least we would get a trip to Hawaii.

10

WHO ORDERED THAT?

If you do not expect the unexpected, you will not find it; for it is hard to be sought out, and difficult.

—Heraclitus of Ephesus (ca. 550–475 B.C.)

JANUARY 9, 1992: Alexander Wolszczan of the Arecibo Observatory in Puerto Rico and Dale Frail of the National Radio Astronomy Observatory in New Mexico published an astonishing result in *Nature*—they stated that they had found strong evidence for the existence of two "planets" in orbit around a *pulsar.*

Pulsars are weirdly exotic stars, leftovers from the death of stars perhaps 10 times as heavy as the Sun. Such massive stars end their lives in supernova explosions, blowing off most of their outer layers and shining as brightly as an entire galaxy of stars for a few months. Massive stars create new elements (e.g., carbon, oxygen, and silicon) in the course of their evolution prior to the explosion and then eject these elements, as well as freshly created elements, during the explosion. Most of these elements end up in interstellar clouds, where they are available to form new stars and planetary systems.

The core of the massive star is compressed into a miniscule star that is so incredibly dense that its protons and electrons are forced together to form neutrons—the entire star is composed of neutrons. A neutron star is thus as dense as the nucleus of an atom, containing the mass of the Sun inside a sphere only a few miles in diameter. If it was any denser, a neutron star would have to become a black hole and disappear from sight, as no light could escape from its surface.

Neutron stars have surface temperatures of as much as a billion degrees Centigrade and immensely strong magnetic fields created by the drastic compression of the progenitor star's relatively weak magnetic field. Neutron stars also tend to rotate extremely rapidly, as fast as a thousand times per second, because the massive star's core must spin faster and faster as it is compressed (similar to a twirling ice skater pulling in his or her arms). The formidable magnetic field of a neutron star channels its radiation into narrow directions, like a searchlight beam, and rapid rotation of the star makes this searchlight beam seem to pulse on and off as it sweeps past the Earth. A pulsar is thus a neutron star that gives off a steady stream of pulses of radio waves, the characteristic phenomenon through which pulsars were first detected and for which they are named.

Pulsars were such unexpected objects that when they were first found in 1967, there was some suspicion that their periodic radio signals might be coming not from a celestial body, but from an intelligent civilization that was intentionally broadcasting its proud existence to the universe. The first pulsar was provisionally labeled LGM, for "little green men"—even though it was found by a woman, Cambridge astronomer Jocelyn Bell. Other pulsars were soon found, however, distributed around the Milky Way galaxy, weakening the case for the LGM being a signal from a uniquely advanced, galactic civilization. In addition, there was no evidence of the Doppler shift expected if the signal arose from a planet orbiting a distant star. The LGM hypothesis was thereafter forgotten, but not the idea that pulsars might harbor planets.

Wolszczan and Frail's announcement came as a surprise. Because pulsars were believed to be formed after a supernova explosion obliterated the outer layers of the original star, a pulsar was not expected to be the most promising place to find another solar system. A powerful supernova explosion would suddenly remove most of the mass of the original star, leaving any planets in orbit without enough of a central star to hold them on their orbits—they would quickly fly off and disappear as a result. So how could there possibly be planets around a pulsar? Maybe the planets were not there at all.

Wolszczan and Frail's announcement was not the first time that planets had been claimed to exist around a pulsar. Soon after the first pulsar was discovered, a team led by D. W. Richards of the Arecibo Observatory found a signal from the Crab Nebula pulsar that they thought might imply the presence of a planetary companion.

The Crab Nebula is the remnant of a supernova that exploded relatively near the Earth in the year 1054, producing a transient brightening that was noticed and recorded by Chinese astronomers almost a millennium ago. The Crab Nebula pulsar can be seen with an optical telescope to be blinking on and off at the same rapid rate (30 times per second) that the radio waves pulsate. These periodic emissions from the pulsar are caused by the rapid rotation of the neutron star itself, once every 33 thousandths of a second. Because the pulsar's pulse is tied to the neutron star's rate of rotation, which changes only very slowly, the periodic signals from a pulsar are extraordinarily regular and clocklike.

Richards and his team followed the Crab Nebula pulsar with the valley-sized (305-meter) Arecibo radio dish for several months, patiently amassing data on its pulsations. After accounting for all known sources of variations, they found that there was still an unexplained periodic signal in their data, something that made the time at which the pulse of radio waves arrived vary back and forth every 77 days. The team concluded that something was causing the pulsar to oscillate every 77 days, possibly an orbiting companion. If the neutron star was forced to orbit around in a tight circle because of the gravity of an unseen companion, then the same Doppler effect that periodically redshifts and blueshifts the lines in a binary star's optical spectrum would also shift the arrival times of the pulsar's radio waves. Richards and his group thus suggested in April 1970 that they had evidence for at least a 1 Earth-mass planet with an orbit about the pulsar less than half the size of the Earth's orbit around the Sun.

However, further observations over a longer period of time showed that the 77-day oscillation was not real. There was no definite periodicity of 77 days in the Crab Nebula pulsar's radio signals—Richards and his group had been misled by noise in the timing system and by their own enthusiasm.

Wolszczan and Frail were trodding on dangerous ground, but they were following in the tracks of yet another group. In a July 25, 1991 letter to *Nature*, Andrew Lyne and his colleagues at Britain's famed Jodrell Bank radio telescope claimed to have detected a 10 Earth-mass planet in a circular orbit around the pulsar PSR1829−10. Pulsar names are formed by joining the designation PSR (for pulsar) with the coordinates of the pulsar on the sky in the reference system used by astronomers— given just the name, then, you also know where the object is located.

The Jodrell Bank group had found 40 new pulsars in a long-term survey, and PSR1829−10 was the only one of the 40 that seemed peculiar. The time of arrival of its pulses seemed to be periodic, changing a minute amount every six months, and Lyne's group had run out of explanations for this periodicity—they thought there must be a planet orbiting PSR1829−10 and causing the variations. Because none of the several hundred other pulsars they had analyzed by the same method showed similar variations, Lyne and his colleagues were fairly certain that they had not introduced any spurious errors into the data for PSR1829−10. PSR1829−10 was definitely acting odd, as if it had come down with a planet.

Lyne's paper was accompanied in the same issue of *Nature* by a commentary by David Black. Black took the discovery seriously but cautiously pointed out that the PSR1829−10 interpretation needed to be confirmed by another group before it could be considered as established. Black also suggested that the most promising way to account for the existence of a pulsar planet was to show that it had been made during the same process that formed the neutron star; perhaps the supernova left behind not only a neutron star, but also a disk from which a planet might form.

Theorists rushed to explain the existence of Lyne's pulsar planet; a flurry of papers was soon published in *Nature* and elsewhere. Because the pulsar planet's orbit was inferred to be circular, it was unlikely to have predated the supernova explosion, which would leave any surviving planet on a very noncircular orbit at best. Hence a means had to be found to either make or capture a planet around a neutron star.

One group suggested that the latter could be accomplished by having a neutron star collide with and consume a solar-type star, acquiring the victim's planetary system to boot, an extreme case of cosmic carjacking. However, most theorists quickly settled on a more likely hypothesis—the neutron star must have gained its planet in the process of eroding to death a stellar companion. The intense radiation (mostly highly energetic particles) given off by a neutron star blows the socks off of anything nearby, including any unfortunate binary companions. One pulsar, PSR1957+20, known as the Black Widow pulsar for its deadly habits, had been shown to be slowly destroying its trapped prey. The Black Widow had already succeeded in reducing the companion's mass down to the range reserved for brown dwarf stars, about 25 Jupiter masses, making the hapless companion a pulsar brown dwarf. Some of the leftovers from this meal might form a disk in which a planet could form.

It was onto this stage that Wolszczan and Frail threw *their* discovery. Like Richards two decades before, they had used the Arecibo radio telescope to monitor their pulsar. Wolszczan had also used Arecibo in 1990 to discover the pulsar, PSR1257+12, lying 1,500 light years away from the Sun. PSR1257+12 had a rapid rotation rate, spinning over a hundred times every second, much faster than PSR1829−10's three rotations per second. Wolszczan and Frail found good evidence that PRS1257+12 was surrounded not by one, but by *two* planets with masses a few times that of the Earth, moving with orbital periods of 98 and 67 days and corresponding to distances from the pulsar a bit less than half the Earth-Sun distance.

Combined with Lyne's result for PSR1829−10, Wolszczan and Frail's claims made it suddenly seem that pulsars had more planets than normal stars like the Sun. After all, there weren't any viable planetary candidates at this time, even of Jupiter mass, much less with the few to 10 Earth masses deduced for the objects found by the radio astronomers.

We were being presented with a universe in which planets were commonplace around pulsars but apparently rare around normal stars. It was unsettling and unnerving. The radio astronomers claimed to have won the grand prize, to have found the first "planets" outside the solar system. Even if they were real, however, the pulsar planets had to be very different from the planets in our Solar System because of the extremely hostile environment in which they formed and orbited—the harsh radiation field of a neutron star would not lead to a hospitable planet like Earth.

JANUARY 15, 1992: Andrew Lyne reported at the American Astronomical Society meeting in Atlanta that his group's claim the year before for having found a 10 Earth-mass planet orbiting the pulsar PSR1829−10 was incorrect. He apologized profusely for the embarassing situation.

The Jodrell Bank team had inadvertently rediscovered the fact that the Earth revolves around the Sun. The source of the six-month periodicity that the team thought might be a planet turned out to be a planet all right—but it was just the Earth, which takes six months to go from one end of its orbit to the other. This known periodic effect was subtracted from the data before the group began its search for planets, but the subtraction assumed that the Earth's orbit was exactly circular, which is usually a good assumption. The Earth's orbit has an eccentricity of 0.02, not zero, however, and that small difference was enough to mislead Lyne and his group. Van de Kamp had joked that he once rediscovered Jupiter by neglecting to subtract from his astrometric data a 12-year variation induced by Jupiter's presence in the Solar System.

I read Lyne's startling admission the next day in a copy of the *Chicago Tribune* that I had picked up in O'Hare Airport. I was delayed at O'Hare by an aircraft mechanical problem while on my way to Hawaii for the TOPSSWG writing workshop, and I needed a break from reading the draft chapters of the TOPS report. I carried the hot news with me to the TOPS meeting, where Bernard Burke, well plugged into the radio astronomy world, already knew about the unfortunate outcome for PSR1829−10's planet.

JANUARY 17–22, 1992: The TOPSSWG met in the Ritz-Carlton Hotel on the Kohala coast of the Big Island of Hawaii, in the morning shadows of the Mauna Kea volcano. The Kohala coast is a desolate, treeless landscape covered by fresh lava flows but embedded here and there with exquisite resorts such as the one we were at, the Mauna Lani. The sun shone brightly each day, reflecting off the waves of the Pacific and the white domes of the observatories high on Mauna Kea, but it might just as well have been raining. We TOPSSWG members were trapped inside the hotel each day, debating each other and writing the text of our report. While we were paying the government room rate (well less than one third the normal rate), we could hardly afford to eat at the hotel on our limited travel budget, so we gorged ourselves on the fresh pastries and juices provided in our meeting rooms.

We had to write a report by the end of the next month that could be used to persuade Congress and our colleagues that the search for extrasolar planets made sense. The sticky issue of the pulsar planets would have to be addressed. Lyne's claim could now be safely ignored, but Wolszczan and Frail's claim was not going to be so easy to handle—there was a good chance they really had something. The TOPSSWG was worried that the impetus of looking for planets around Sun-like stars might somehow lose momentum because of the discovery of planetary-mass objects around pulsars. It was unclear if we should even dignify these poor cinders with the name "planets," with or without quotation marks. The TOPSSWG believed that the ultimate goal was to find planetary systems similar to our own, capable of harboring life and sharing a common history, and certainly the pulsar planets failed to meet any of those criteria. We decided that while we had to acknowledge the possibility of their existence, though they were as yet unconfirmed, we should not let the pulsar planets derail our efforts to launch the TOPS program.

We wrote more text for the report on laptop computers and argued in plenary sessions and splinter groups for four and a half days. In between we had one day to take a tour of the Keck Observatory on Mauna Kea. Going from sea level to almost 14,000 feet in a few hours of driving tends to be a bit rough on most people, so we stopped for lunch at the 9,000-foot level at Hale Pohaku, on the flank of Mauna Kea, where the astronomers sleep each day in preparation for the night's observing. After acclimating to high altitude for two hours at Hale Pohaku, we ascended in four-wheel-drive vehicles to the summit, passing the growing array of optical and millimeter-wave telescopes that is beginning to crowd the dormant volcano's top. (In addition to the Keck Observatory, NASA's IRTF, the Canada-France-Hawaii, and other optical telescopes, both Caltech and the United Kingdom had built large millimeter-wave telescopes on Mauna Kea. By the time that a few more telescopes are built, the mountaintop will have reached the maximum number of telescopes permitted by environmental impact concerns.) The summit panorama stretches from the Mauna Loa volcano to the south, to the Haleakela Crater, which is vaguely seen about 80 miles away on the neighboring island of Maui.

The first Keck telescope was nearing completion, with about half of its complement of 36 hexagonal-shaped mirrors already in place. We climbed up on the Keck telescope's superstructure to the instrument platform for the best view down at the mirrors. A 10-meter-wide telescope is an impressive object indeed, especially considering the humble diameters (a few feet) of the telescopes that had borne the brunt of the searches for extrasolar planets so far. The TOPS program was going big time with the Keck involvement.

One person from our group did not enjoy the tour as much as the rest of us: Mike Shao dropped out and went to sit in one of the four-wheel-drive vehicles. I knew exactly how he felt—when my family took the cog railway to the top of Colorado's 14,110-foot-high Pike's Peak on a vacation in the 1960s, I foolishly got off the train and ran excitedly around. Within minutes I was sitting with my head down, feeling distinctly uneasy and nauseous, a feeling that did not go away until we had descended the mountain.

Altitude sickness is common on 13,796-foot-high Mauna Kea, Hawaii's tallest peak, and just about everyone's abilities are impaired to some degree by the shortage of oxygen. The engineers who built the telescopes on Mauna Kea soon learned to make a detailed list while at sea level of what they intended to accomplish on the mountaintop. Once at the summit, they followed their list and didn't expect to do any creative thinking or problem solving, because they would probably fail. For the same reasons, astronomers are warming to the idea of doing their observations by remote control, while sitting at the Observatory's headquarters in low-lying Waimea.

MARCH 26, 1992: The TOPSSWG met in Washington to hear the status of the TOPS report and program. Wesley Huntress told us that NASA had failed to get Congress to put in any new money to pay for the Keck Telescope in the coming year's budget, but that NASA would try again for the next year. Unfortunately, the next year's battle for new money was expected to be even harder than the one NASA had just lost. The Keck people were being patient and would wait for NASA to come up with the money, but only up to a point. If NASA did not succeed in the next budget cycle, the Keck Observatory would have to start looking for a new partner.

The TOPS report was in somewhat better shape than the effort to pay for NASA's share of Keck, with the final revisions underway and with plans to print 10,000 copies. Maybe putting the TOPS report out on the street would help NASA convince Congress of the importance of finding new solar systems.

APRIL 1992: President George Bush appointed Daniel Goldin to be the new Administrator of NASA, with the blessing of Senator Albert Gore, leader of the Senate's Democrats. Goldin was thus appointed with clear bipartisan support that was likely to survive any outcome of the presidential election later in the year.

The astronaut was out. Goldin was in, and it looked like he would be in for quite some time. What would this mean for NASA? What would it mean for NASA's science programs? And what about the nascent search for other Earths? It would all depend on Daniel Goldin and his interests. When people say that NASA has decided something, what they usually mean is that the NASA Administrator, or one of his close associates, has made a decision. Goldin's support would be crucial to an endeavor as expensive and high profile as the search for Earths.

Goldin was a mechanical engineer by training who had landed his first job at NASA's Lewis Research Center in Cleveland, Ohio and then moved to the corporate world of Thompson-Ramo-Wooldridge (TRW) in California. Goldin's background in building secret surveillance satellites at TRW for the Department of Defense suggested he might be supportive of NASA's robotic science missions, but no one knew for sure.

Goldin's penchant for wearing cowboy boots along with his business suits suggested a maverick personality. If so, NASA Headquarters would be in for a wild ride.

MAY 19, 1992: Gamma Cephei's possible planetary companion, announced in 1988, fell prey to the great danger of continuing to take data in order to improve results.

In a paper submitted to the *Astrophysical Journal,* Gordon Walker disavowed his group's earlier claim for spectroscopic evidence of a 1.7 Jupiter-mass planet orbiting around the star Gamma Cephei. The additional four years of data gathered by the group now implied an orbital period for the planet of 2.52 years, suspiciously equal to the period at which spectral lines in Gamma Cephei's spectrum varied. The spectral line variation indicated that Gamma Cephei, a yellow giant star much larger than the Sun, probably rotated with a period of 2.52 years. The coincidence of these two periods was strong evidence that the star's rotation was causing the spurious velocity signal—apparently Walker's group had been measuring velocities associated with the upwelling and downwelling of gases in the outer layers of the giant star, not an orbiting planet.

By the time of the 1992 retraction, Bruce Campbell, the leading member of the original Canadian team, had quit the field of astronomy, thoroughly discouraged by the difficulty of securing a permanent position as an astronomer in Canada. The difficulty of finding planets evidently did not help.

JUNE 6, 1992: Brown dwarfs were still getting into trouble, too. A team of University of Arizona astronomers led by Todd Henry submitted a paper to *Astronomy and Astrophysics* reporting their studies of the Wolf 424 binary system that Wulff Heintz had found to be composed of two brown dwarf stars with masses of 50 and 60 Jupiter masses. Henry's group had used McCarthy's speckle interferometer (of VB8 B fame) as well as a new speckle camera on the Smithsonian's large 6.86-meter Multiple Mirror Telescope to follow Wolf 424 for most of the last 10 years.

Henry's team found that Wolf 424 was not behaving in the way it should if it was composed of two brown dwarf stars. Instead, the binary was revolving around its center of mass significantly faster than it should be. If the binary stars were moving faster than they should be, at a given separation, then there must be much more mass in the system in order to have enough gravity to balance out the centrifugal forces coming from the faster orbital speeds. Wolf 424 could not then be a pair of brown dwarf stars, but rather a pair of garden variety, faint low-mass stars.

This conclusion was supported by other observations by Henry's group of the spectrum and brightness of Wolf 424 at infrared wavelengths. The infrared spectrum of Wolf 424 was a near-perfect match to that of a standard low-mass star, showing multiple bands of absorption

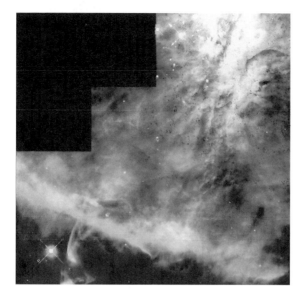

Hubble Space Telescope image of the Orion Nebula, a nearby interstellar cloud filled with bright young stars (stars with X-shaped diffraction features) and with stars in the process of formation. Small clumps of gas and dust in which new stars are forming can be seen in front of the background nebulosity. (Courtesy of C. Robert O'Dell, Rice University, and NASA.)

Hubble Space Telescope image of the Eagle Nebula, a more distant star-forming region where ultraviolet light from bright stars is eroding several "elephant trunks" of gas and dust on the edge of a dark cloud. Several of the trunks appear to contain young stars in the process of formation. (Courtesy of Jeff Hester and Paul Scowen, Arizona State University, and NASA.)

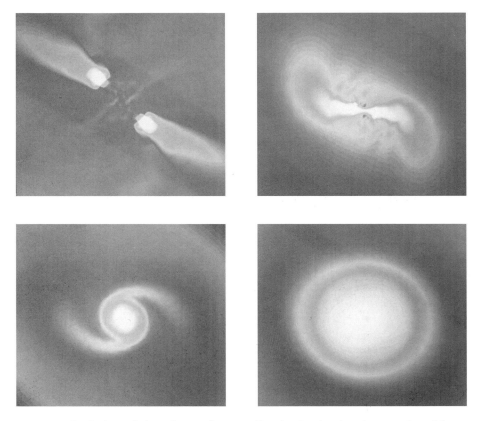

Computer simulations of the collapse of protostellar clouds, showing the cessation of fragmentation into smaller mass protostars as the mass of the cloud is lowered. White represents highest density. The four clouds are identical except for their initial masses: 2, 0.25, 0.1, and 0.02 times the mass of the Sun. (Calculations by Alan Boss, Carnegie Institution of Washington.)

Alan Boss (left) and George Wetherill (right) in the library of the Carnegie Institution of Washington's Department of Terrestrial Magnetism. Wetherill is the leading worker on the theory of forming Earth-like planets. (Courtesy of Alan Boss.)

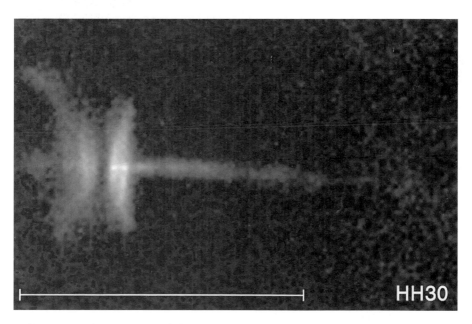

Hubble Space Telescope image of the suspected protoplanetary disk around the young star associated with the HH-30 outflow in Taurus. The outflow (horizontal streak) points back to the nearly edge-on disk (dark area between bright arcs) hiding the young star (unseen). The bright arcs are caused by starlight scattered off the surface of the disk, which can thus be seen to be curved. Scale bar is 93 billion miles. (Courtesy of Christopher Burrows, Space Telescope Science Institute, and NASA.)

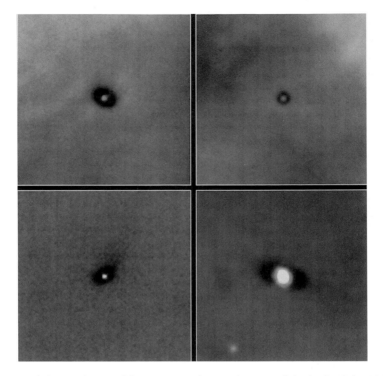

Hubble Space Telescope image of four suspected protoplanetary disks in the Orion Nebula. The disks are seen in silhouette in front of hot, ionized gas and have sizes about two to eight times the size of our Solar System. (Courtesy of Mark McCaughrean, Max Planck Institute for Astronomy, C. Robert O'Dell, Rice University, Space Telescope Science Institute, and NASA.)

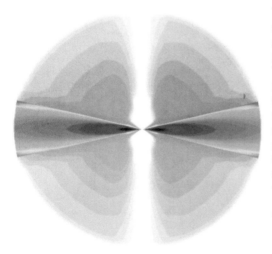

Theoretical model of a planet-forming disk, seen in cross section, with black corresponding to the highest-density regions in the midplane of the disk. The region shown is 10 times the Earth-Sun distance in radius; the protosun (unseen on this scale) lies at the center. The disk rotates about an axis that is a vertical line through the center of the disk. The low-density regions along the rotation axis provide a bipolar funnel for the escape of the protosun's wind. (Calculation by Alan Boss, Carnegie Institution of Washington.)

Michel Mayor (left) and Didier Queloz (right) of the Geneva Observatory, discoverers of the first extrasolar planet around a star like our sun, 51 Pegasi. (Photograph by H. Pignon, courtesy of M. Mayor.)

R. Paul Butler (left) and Geoffrey Marcy (right), of San Francisco State University, discoverers of the first giant planet on a "normal" orbit around a solar-type star, 47 Ursae Majoris. (Courtesy of San Francisco State University.)

Hubble Space Telescope image of the first cool brown dwarf star, the companion to Gliese 229. The brown dwarf is the small white object, orbiting about 44 times the Earth-Sun distance from Gliese 229, whose overpowering light caused the bright halo and diffraction spike jutting diagonally across the image. (Courtesy of the Space Telescope Science Institute and NASA.)

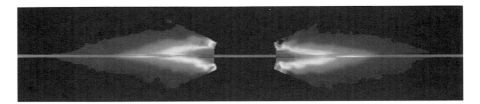

Hubble Space Telescope image of the dust disk surrounding the nearby star Beta Pictoris. In this image, a slight warp of the innermost disk is evident (asymmetry in the innermost, dark region), possibly caused by an unseen giant planet. Beta Pictoris itself has been removed from the image, allowing the faint disk to be seen on a scale of a hundred times the Earth-Sun distance. (Courtesy of Christopher Burrows, Space Telescope Science Institute, and NASA.)

Computer model of the formation of a gas giant proto-planet by the gravitational instability of a protoplanetary disk. The protoplanet (small bright streak at twelve o'clock) is a multiple-Jupiter-mass clump of gas and dust that orbits in a spiral arm at about eight times the Earth-Sun distance from the central protosun (unseen). (Calculation by Alan Boss, Carnegie Institution of Washington.)

Artist's illustration of the interferometer planned by NASA for construction at the Keck Observatory on Mauna Kea. Four outrigger telescopes of 2-meter size would be combined with the two 10-meter-diameter Keck telescopes to produce a powerful interferometer, capable of detecting Jupiter- and Neptune-mass planets around nearby stars by the astrometric technique. (Courtesy of JPL and NASA.)

Artist's conception of the Space Interferometry Mission (SIM) planned by NASA for launch into space around the year 2005. SIM will be an optical interferometer about 10 meters in length, with the ability to detect Neptune-mass and perhaps Earth-mass planets around nearby stars by the astrometric technique. (© 1998 California Institute of Technology. All rights reserved.)

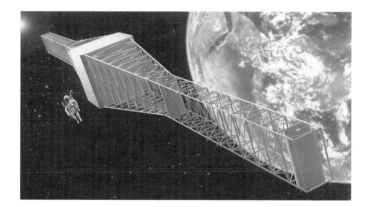

One concept for the Planet Finder, an infrared interferometer planned by NASA and capable of directly detecting Earth-mass planets and characterizing their atmospheres. The Planet Finder would be about 75 meters in length and would have four 1.5-meter telescopes arranged as a pair of interferometers that cancels out the light from the planet's star, allowing the planet's light to be imaged. Planet Finder may have to travel to Jupiter's orbit to avoid the infrared emission from the inner Solar System's zodiacal dust cloud. (Courtesy of Roger Angel, University of Arizona.)

Simulated appearance of an extrasolar planetary system, as imaged in the infrared by the Planet Finder. The hypothetical system consists of four planets, each of which is easily seen after just 10 hours of observing. (Courtesy of Roger Angel, University of Arizona.)

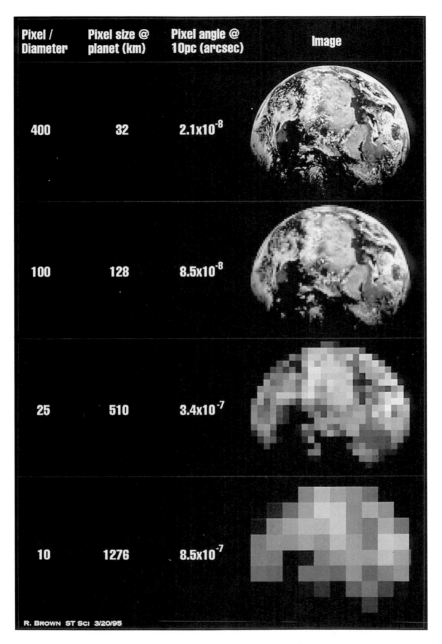

Pixel / Diameter	Pixel size @ planet (km)	Pixel angle @ 10pc (arcsec)	Image
400	32	2.1×10^{-8}	
100	128	8.5×10^{-8}	
25	510	3.4×10^{-7}	
10	1276	8.5×10^{-7}	

R. BROWN ST SCI 3/20/95

The Earth, as imaged by a telescope at progressively greater distances, showing that hints of continents and clouds can be discerned even with just 10 pixels (resolution elements) across the planet's diameter. Achieving this (or even higher) resolution for images of extrasolar planets will be a severe challenge for future astronomers. (Courtesy of Robert A. Brown, Space Telescope Science Institute.)

from molecules such as titanium oxide. The brightness of Wolf 424 was about what was to be expected if it was composed of two 100 Jupiter-mass stars, making the two interestingly low-mass objects, but not brown dwarfs.

With the Texas group having shot down George (HD 114762) the year before, it seemed that two more brown dwarfs had bit the dust of the American Southwest.

JULY 2, 1992: In the midst of these grim days for extrasolar planet and brown dwarf searches, at long last there emerged the first *confirmation* of a claimed discovery of planets outside the Solar System. Unfortunately for those of us hoping for solar system analogs, the long-anticipated confirmation came from radio astronomers.

Wolszczan and Frail's bold claim at the beginning of the year for two multiple-Earth-mass objects in orbit about the pulsar PSR1257+12 was splendidly confirmed by an independent group of astronomers. UC Berkeley's D. Backer and his colleagues used the National Radio Astronomy Observatory's 140-foot radio telescope in Greenbank, West Virginia to monitor the pulsar. After analyzing the observations with two different computer packages, the Berkeley group saw no reason to doubt Wolszczan and Frail's interpretation.

Pulsar PSR1257+12 really was orbited by two objects with masses a few times that of Earth. Even if the lack of beach-front property on those barren worlds meant that real estate agents would not push for NASA to colonize PSR1257+12's planets, the fact that Earth-mass objects could form in the hostile aftermath of a supernova explosion meant that the planet formation process had to be remarkably robust, not a finely tuned and synchronized choreography that could only take place under special circumstances. If pulsars had planets, then surely so did normal stars like the Sun, if only we could just *find* one.

OCTOBER 5, 1992: The TOPSSWG met in Washington to react to a proposal by NASA's managers to try to protect the SETI program by placing SETI inside the TOPS program and giving it a new name. This was somewhat like trying to protect the life of a star witness in a high-stakes criminal case through a quick change of identity and a move to another state.

The search for extraterrestrial intelligence had always aroused strong feelings in people, either positive or negative, and NASA's SETI program was about to embark on a public phase of existence. On the symbolic date of October 12, 1992, five hundred years after Columbus arrived in the New World, NASA's SETI project was going to be launched with appropriate fanfare: SETI's scientists would begin a powerful new search using electronic receivers specially designed to sift through radio-wavelength noise in search of messages sent from other worlds. At the same time, SETI's critics in Congress were fuming at the idea of NASA spending 12 million dollars each year on a search for little green men and women. SETI's Congressional enemies wanted blood: 46 million dollars had already been spent in developing the specialized receivers and planning the search, and some said that was enough.

Everything was coming to a head, and NASA decided it would be best to rename SETI as the High Resolution Microwave Survey project, or HRMS for short. The new name was scientifically correct, if a bit misleading—who would think from the name alone that the purpose of the survey was to listen at microwave wavelengths for signals from another civilization in the galaxy?

In addition, HRMS would become a part of the TOPS program—in fact, the *only* part of the TOPS program with a real budget, given that the TOPS plan itself, including the commitment to Keck, was still not in the NASA budget. Without a budget entry, a program does not exist, and the TOPSSWG was now told that *this* was the best way to get the TOPS program started, not as a new program, but through the back door of becoming affiliated with an existing NASA program with a substantial budget.

The TOPSSWG did not like the idea of the switch at all. We had been notified ahead of time about what was in the wind, and a storm of letters of protest was sent to NASA Headquarters. The TOPSSWG felt there was a real danger that the TOPS/HRMS program would become the next target of SETI's Congressional enemies, and the most important part of TOPS, searching for new solar systems, had not yet made it into the NASA budget. The TOPS planet search program was already in an extremely vulnerable position and did not need any further perils. The protests of the TOPSSWG were to no avail; NASA had already done what it felt it had to do to save SETI.

Congress was not amused by the SETI shell game. One Representative bitterly complained that "maybe TOPS should stand for Taking Ordinary People's Savings."

At the TOPSSWG meeting, Lennard Fisk put the best face on matters by telling us that by virtue of the creation of TOPS/HRMS, the TOPS program was now underway. There was still no official sign about whether the Administration would ask for the Keck Observatory money in the next budget cycle, but we were given the ominous suggestion that the TOPSSWG might have to reconsider the Keck option. We had just had our program's cherished acronym appropriated for another purpose, and now we began to worry that we were about to lose the chance to join the Keck Observatory.

On the positive side, we were told that Daniel Goldin had read the TOPS report and had expressed enthusiasm for the idea of searching for new planets. The only problem, not surprisingly, was finding the money to pay for it.

The question of how best to plan for linking the two Keck telescopes into a single giant optical interferometer also surfaced at the TOPSSWG meeting. JPL, in anticipation of running the show at Keck, wanted to begin to develop its expertise in the area of interferometry by building a testbed in which the basic principles of combining and manipulating optical light beams could be demonstrated and tested. The idea of building a testbed prior to making the huge leap to the Kecks seemed prudent.

JPL's fast-talking Mike Shao presented a plan for building a multi-million-dollar testbed facility on Palomar Mountain, home of the 200-inch Hale Telescope and conveniently located between Los Angeles and San Diego. After his experience on top of Mauna Kea, Shao was content to consider placing the testbed on 6,140-foot-high Palomar mountain, where serious thinking could be done without worrying about wearing an oxygen mask. As the leader of the JPL interferometry effort, Shao's expertise was critical to the project, and if Shao preferred Palomar to Mauna Kea, that was that. Compared to getting to Mauna Kea, the several-hour drive to Palomar from Pasadena would certainly simplify the lives of the JPL engineers who would develop the testbed.

OCTOBER 29, 1992: After several days of meetings at NASA Headquarters, it was always a palpable pleasure to return to my office at DTM and get back to doing my own research. On this day I caught George Wetherill in the elevator at DTM and showed him a plot I had just created. After working for several years to calculate the temperature distribution inside the solar nebula, I finally had a firm result.

Knowing what the temperature was inside the solar nebula is about as important for planet formation theorists as knowing the temperature outside your house—if you go out, are you going to freeze or sweat? Months of computing on a fast workstation led to my prediction that the temperature at the middle of the solar nebula should drop down below the ice condensation temperature (about −120 degrees Centigrade) at a distance from the Sun a little more than four times that of Earth. Solid ice particles could not exist inside this radius; it was simply too hot for them. The result also meant that gas giant planets, which require the presence of solid ice planetesimals in order to grow large enough to capture gas from the nebula, could not form inside this distance—only rocky planets could form there.

The upshot of the calculation was that the innermost giant planet should occur at about five times the Earth-Sun distance from its star. This distance is precisely that of Jupiter's orbit. Only terrestrial planets should be found inside this distance, as is indeed the case for our Solar System. My calculation had no hidden dials to fiddle with in order to make sure that the outcome agreed with the characteristics of our Solar System—it had taken many months just to calculate this single model, so I had to make my best guess about what the mass of the solar nebula should be and at what rate it should be gaining mass, and then let it run, fingers crossed the whole time.

The calculations strongly supported the idea that we understood the basics of how the Solar System formed. I had computed the temperature inside a solar nebula with the characteristics of the protoplanetary disks being studied by astronomers like Sargent and Beckwith, the best known analogs for the solar nebula. The resulting temperatures implied that Earth-like planets should form inside about four times the Earth-Sun distance, and gas giant planets outside this distance. The structure of our Solar System, of course, had been long known, so the result could hardly be considered a prediction; the real test of its validity could only come when *new* solar systems were discovered.

DECEMBER 8–10, 1992:   JPL organized an international conference on the Caltech campus about the search for extrasolar planetary systems, at which George Wetherill presented a scary idea. Wetherill said that there was a good chance that we should not necessarily expect to find gas giant planets like Jupiter in orbit around other stars just because we have such benevolent monsters in our Solar System.

Wetherill was used to either including the gravitational kick of Jupiter in his simulations of the formation of the Earth or to ignoring Jupiter, so he was predisposed to consider what would happen to comets in a solar system without a Jupiter. What he learned was disheartening to the planet searchers at the conference.

If Jupiter did not exist in our Solar System, Wetherill pointed out, we would not be alive today to wonder about it, because his calculations showed that Jupiter played the key role in kicking comets safely out of the Solar System and away from Earth. Without Jupiter looking out for us, the rate at which comets struck the Earth would be about a thousand times higher than it really is. This meant that comets big enough to strike the Earth and wipe out species like dinosaurs (and mammals) would arrive roughly every 100 thousand years, instead of every 100 million years. It was unlikely that a planet being pelted with 10-kilometer-sized comets every 100 thousand years would be able to evolve intelligent life—it took roughly that long for *H. sapiens* just to evolve from its humanlike predecessors. To evolve fully intelligent life from the survivors of a dinosaur-killing impact before the next big one hit seemed highly improbable. A solar system without a Jupiter might be perpetually populated by cockroaches, not E.T.-like aliens.

Wetherill told the shocked audience that we thus could not be sure that giant planets existed elsewhere—in some sense, we did not even have a sample of one (our Solar System) that we could expect to be representative of planetary systems in general. Our Solar System might well be the odd exception rather than the rule.

The fact that no one had yet been able to find any Jupiter-mass planets around nearby stars fully supported Wetherill's hypothesis. The University of Arizona's Robert McMillan and Texas's William Cochran, both of whom had been searching for extrasolar Jupiters since 1987, were put on the spot and asked if they had *any* evidence for Jupiters in their spectroscopic data. McMillan and Cochran both cleared their throats and cautiously stated that they had seen nothing as yet. With their somewhat shy personalities in public, neither was the type to blurt out something that might later have to be retracted.

The sickening possibility had arisen that we had failed to find Jupiters not just because of the inherent technical difficulties, but because they simply were not there to be found. If Jupiters were rare, the task of finding the first extrasolar planet around a normal star became immensely more difficult—we would then have to plan our search to find Uranus- and Neptune-mass planets, objects more than 10 times smaller in mass and hence much harder to find.

11

THE BATTLE OF PALOMAR MOUNTAIN

The best person to decide what research shall be done is the man who is doing the research. The next best is the head of the department. After that you leave the field of best persons and meet increasingly worse groups. The first of these is the research director, who is probably wrong more than half the time. Then comes a committee, which is wrong most of the time.

—Charles Edward Kenneth Mees,
Research Director of Kodak (1882–1960)

FEBRUARY 17–18, 1993: The TOPSSWG met in Washington to consider the plans for turning the Keck telescopes into an interferometer and the prospects for the orbiting planet-search telescope.

At the previous meeting, JPL's Mike Shao had proposed a facility at Palomar Mountain as the first step toward interferometry with the Keck telescopes. In response, this meeting included presentations from several university and government groups that had already been planning or even building optical interferometers that could serve the same purpose as the Palomar testbed. Harold McAlister of Georgia State University had plans for and would soon have NSF support to build an array of optical telescopes that would be linked to form an interferometer with a resolving power on the order of a 100-meter-diameter telescope. The Naval Research Laboratory was already building a six-element interferometer near Flagstaff, Arizona, and another optical interferometer was underway on Arizona's Mount Hopkins. Any one of these facilities could use some extra money and might serve as a precursor to the Keck interferometer.

The presentation of JPL's latest ideas for the Palomar testbed stirred the fires of the TOPSSWG. The university-based scientists could see the TOPS program disappearing whole down the voracious mouth of JPL. A prize like TOPS did not come around often and was worth a fight.

University scientists by and large obtain funds from the federal government through the process of peer review, in which an orderly competition is announced; proposals for research or equipment are solicited, written in great detail, and mailed out for comments and ratings from other scientists; and then, finally, panels of scientists gather to criticize and discuss the proposals and ultimately to choose the winners. The process is arduous and drawn out but hard to improve on if one wishes to allocate precious research funds on the basis of merit (at least merit as perceived by the scientists involved in the review process). Thus university scientists do not expect to be handed a big pot of money to perform a research task—they expect to have to compete for it, and they expect others to have to compete as well.

Accordingly, members of the TOPSSWG argued that an open competition should be held to determine who would win the right to develop Keck interferometry with NASA's support. However, the TOPSSWG was told that NASA Headquarters had already given JPL the "programmatic responsibility" for the TOPS program. JPL badly needed this promising program and the technical jobs that would come along with it; unlike the case at other NASA centers, JPL scientists are not civil servants, and their job security is dependent on a steady flow of federal contracts. NASA helps to support the research of many university scientists, including some of those on the TOPSSWG, through the highly competitive, peer-reviewed research programs, but in the case of the TOPS program, NASA decided to act like a corporation and unilaterally award the work to what is effectively one of its own divisions, rather than to contract the work outside the corporation. That crucial outcome of the battle of Palomar Mountain was decided before the battle was even engaged.

The issue then arose of the best location for the testbed. Given that the ultimate motivation was to link the Keck telescopes into an interferometer, perhaps with the addition of several smaller, "outrigger" telescopes, Mauna Kea seemed like the right place for the testbed. After all, it was not clear if the windswept top of a dormant volcano, adjacent to several active volcanoes, was a suitably stable site for the extremely delicate task of combining optical light beams. It was also not clear if the Keck telescopes, with their innovative design involving 36 mirrors linked by computer control into a single coherent mirror, might have trouble preserving the extreme stability needed for interferometry. If a testbed had to be built, several members of the TOPSSWG thus argued, it should at least be built on Mauna Kea.

With the TOPS report completed and released, Bernard Burke stepped down as chair of the TOPSSWG halfway through the meeting and was succeeded as chair by David Black. NASA Headquarters planned that Black would also head a committee to review JPL's proposal for the Palomar testbed, with the review committee including members of the other three competing optical interferometers. The TOPSSWG could only hope that the Palomar testbed would receive the appropriate degree of scrutiny from the planned review committee.

NASA's space science budget was under such intense pressure, we were told, that it did not make sense to try to downselect one of the three concepts for the orbiting planet-search telescope. The start date for such a telescope was receding farther into the future faster than the future was approaching, and it seemed premature to pick the winning team so early in the game. All three concepts remained alive—the now combined ATF/CIT, OSI, and POINTS—as well as a fourth idea from NASA Ames scientist William Borucki: FRESIP, which stands for Frequency of Earth-sized Inner Planets. FRESIP was intended to find evidence for planets as small as Earth by watching a large number of stars and waiting for the brightness of one star to decrease a small amount when an orbiting planet passed in front of the star and blocked some of its light.

The anticipated fight to the finish for the orbiting telescope was delayed again—the Astrophysics Division had not yet chosen between OSI and POINTS for the AIM mission.

Though no formal mention was made of the fact at the TOPSSWG meeting, just a few weeks before Robert Harrington had died unexpectedly of throat cancer at the age of 50. Through his astrometric work on Barnard's star and VB8, Harrington had been one of the pioneers of the search for extrasolar planets and brown dwarfs. Harrington had also been one of the pioneers of using computers to study the stability of the orbits of triple star systems and of planets in binaries. He could often be found at any hour of the day or night, any day of the year, loading calculations into the Naval Observatory's computers. Harrington seemed perfectly healthy and happy at the April 1992 International Astronomical Union meeting on binary stars held near Atlanta, Georgia and organized by McAlister. Less than a year later, Harrington was gone.

Daniel Goldin was in the process of reshaping and redirecting his agency. A few weeks after the TOPSSWG meeting, Wesley Huntress became the head of NASA's renamed Office of Space Science, replacing Lennard Fisk. Huntress would be responsible for prioritizing all of NASA's science programs, including the TOPS program. Given that Huntress had previously been the head of the Solar System Exploration Division, where the TOPS program had been born and championed, Huntress's promotion seemed auspicious to those of us who wanted NASA's help in finding new planets.

JULY 30, 1993:   John Stauffer and his colleagues submitted a paper to the *Astronomical Journal* containing new evidence for several brown dwarf stars. Instead of looking for the reclusive prey in orbit around normal stars, which had been singularly unsuccessful so far, Stauffer decided to try the approach attempted by William Forrest in the Taurus molecular cloud. Stauffer and his team would look for free-floating brown dwarf stars, not gravitationally bound to any other stars—single stars out on their own, just like our Sun.

Stauffer knew that finding a free-floating brown dwarf in the Sun's neighborhood would be extraordinarily difficult, because it could lie in any direction. It would also be incredibly faint, because if it was as old as the Sun, it would have cooled down so much as to become nearly invisible. Instead, Stauffer decided to look in a much more promising location, the nearby Pleiades cluster of stars, a little over 400 light years away and containing over 500 stars only 70 million years old. Brown dwarf stars in the Pleiades should have the same ages as the other stars in the cluster, which are over 60 times younger than the Sun, and should shine about a hundred times brighter as a result. While challenging, searching the Pleiades seemed to be as promising as any other approach.

Stauffer and his colleagues used the 200-inch Hale Telescope to image the Pleiades in two different wavelength bands, the visible and the near infrared. A cool brown dwarf would give off considerably more light in the near infrared than at the shorter visible wavelengths, so Stauffer could search the images for promising candidates by finding the stars that were unexpectedly brighter in the infrared image than would be expected based on the visible light image.

Stauffer's team found six good candidates. The faintest one of all, called Palomar Pleiades 15 (PPl 15), was estimated to have a mass of about 60 Jupiters, based on its brightness and assumed age. Because it was not in a binary system, PPl 15 could not be weighed in the usual manner by determining its orbit around another object, so the mass estimate could not be confirmed. Without more information about PPl 15 and its cohort, it could not be proven that they were really brown dwarf stars; they could be old, faint, low-mass stars that by chance happened to be lying in the direction of the Pleiades. Given the rotten luck of previous search efforts, there was no particular reason to be terribly optimistic about PPl 15.

There was another way, though, to tell if PPl 15 was really a brown dwarf. A group of Spanish astronomers and astrophysicists from the Canary Islands, site of the observatory that is Europe's astronomical answer to Mauna Kea, had proposed the *lithium test* in 1992. Lithium is a very light atom, the next heaviest after hydrogen and helium. Like hydrogen and helium, lithium was produced in the Big Bang that created the universe, though only in trace amounts. Nevertheless, by virtue of the presence of lithium at the beginning, every star should start its life with a bit of lithium. However, once a star becomes hot enough to undergo hydrogen fusion, its lithium will also be consumed by nuclear reactions—a temperature of only a few million degrees Centigrade is all that is needed. Any brown dwarf with a mass of 70 Jupiters or less, though, should never get hot enough to burn its lithium.

Rafael Rebolo and his group at the Instituto de Astrofísica de Canarias thus proposed that brown dwarf candidates be given the lithium test—if they showed evidence of having lithium in their atmospheres, by virtue of having decreased emission at the wavelength predicted for absorption by lithium atoms, then they were genuine brown dwarf stars.

Somebody would have to take another look at PPl 15 and give it the lithium test.

OCTOBER 1, 1993:   The new fiscal year began for the U.S. government, but something was missing from NASA's budget. SETI's Congressional enemies had finally succeeded in eliminating NASA's search for extraterrestrial intelligence, after two decades of careful planning and a single year of operation. The year-old name change from SETI to TOPS/HRMS had not helped one bit. The one element of the TOPS program (HRMS) that had been in the budget was now no longer in the budget; the seed for the growth of the planet-search portion of the TOPS program no longer existed.

But Congress had not killed the SETI project itself. Frank Drake, who conducted the first search for alien radio waves in 1960 using an 85-foot telescope in West Virginia, was the head of the SETI project and was not about to give up on a three-decade effort. Drake, Jill Tarter, and their colleagues scrambled to find private support for the visionary project. SETI's headquarters are located near NASA Ames Research Center in the middle of California's Silicon Valley, and the SETI project succeeded in raising enough money from Silicon Valley entrepreneurs and others to keep the search alive, albeit with a fittingly new name—Project Phoenix, named for the mythical bird that arises anew from its own ashes.

DECEMBER 13-15, 1993: The second international conference on detecting extrasolar planets was held at the Waikoloa resort on the Big Island of Hawaii, close to the Mauna Lani resort, where the TOPSSWG finished writing the TOPS report. David Black told me that after the TOPS/HRMS debacle, the TOPS acronym meant certain death on Capitol Hill. From now on, the TOPSSWG would have to go back to using its old name, the PSSWG, and a new acronym would have to be found for the TOPS program.

George Gatewood reported on the progress of his attempts to detect extrasolar Jupiters using his astrometric device on Pittsburgh's Thaw refractor. Barnard's star showed no evidence of a wobble caused by any roughly Jupiter-mass planet with a period less than about six years; neither did Epsilon Eridani, which by now had a history of claims and retractions almost as long as Barnard's star did. However, another star showed a hint of a wobble—Groombridge 1618, a nearby star that might have a pair of brown dwarfs in orbit, if the early trends were maintained. Only time would tell, and Gatewood, like van de Kamp, was a patient, deliberate man.

FEBRUARY 28-29, 1994: The "Piss-swig" was back, and we met again at David Black's Lunar and Planetary Institute, located this time in a sparkling new building Black helped to design, next to the Johnson Space Center. The ceiling of the building's circular Great Room contained spotlights arranged in an asymmetric pattern familiar to backyard astronomers as the constellation Taurus, Black's astrological birth sign. The Director also had his own parking space in the front of the building, where Black's Lexus was parked on the days he did not ride his bicycle into work.

The PSSWG was relieved to learn that NASA was at last beginning the process of paying for its one-sixth share of the two Keck telescopes, thereby meeting the deadline imposed by the Keck Observatory. NASA was unable to convince Congress to provide new funds for the Keck involvement, though, and so would have to scrape up the money from within its existing budget. The magical attraction of searching for new planets had failed to loosen Congress's ever-tightening grip on the purse strings, even with the formidable arguments made for the search in the TOPS report. The SETI name game had not helped.

By 1994 the Kecks' price had risen (through inflation) to over 40 million dollars, and NASA had struck a deal with the Keck Observatory to pay for its share on the convenient installment plan—yearly payments of 6.8 million dollars through the year 2000. But there was no money available to pay for any new instruments for the Keck, such as specialized devices to search for planets; nor was there money to begin turning the Keck telescopes into a giant optical interferometer, as envisioned in the TOPS report. By the time NASA paid the 6.8 million a year to pay for its share of building the telescopes, plus 2.5 million each year to help pay for the operations at the Observatory, NASA felt it could promise to do no more.

The PSSWG decided that this state of affairs made no sense. They argued that buying into the Keck Observatory and then not providing for building special-purpose planet-detecting instruments was like building a robotic spacecraft and launching it off toward a Solar System target without putting any scientific instruments on board. The sneaking suspicion among scientists was that JPL's engineers would be all too happy to do just that.

The Keck Observatory would come equipped with a number of state-of-the-art instruments, but the PSSWG still felt there was a need for more. In particular, the Keck I telescope would have a spectrometer (called HIRES, for High-Resolution Echelle Spectrometer) that could be used for searching for planets by the spectroscopic method, but it seemed to some people to be a waste of valuable Keck time to use HIRES for this purpose when existing high-precision spectroscopic programs using moderate-sized telescopes, such as Cochran's, McMillan's, and Marcy's, were already underway. One idea for a specialized Keck instrument was a coronagraphic camera, a Keck-sized version of the device Smith and Terrile had used to image the Beta Pictoris disk.

Meanwhile, the Palomar testbed was ready to enter the construction phase on the mountain, supported by 3 million dollars in NASA funding. David Black's review committee had met, but it was clear that the Palomar testbed was a fait accompli: JPL had the green light from NASA Headquarters.

The facility was intended to test an idea for doing astrometry with the Kecks that promised a tremendous improvement in precision—the idea of narrow-angle astrometry. If a suitably bright, distant star was located no more than a small angle away from the target star, then both stars would be distorted in the same way by the troublesome turbulent bubbles in the upper atmosphere. The planet-induced wobble of the target star could then be detected by measuring the position of the target star with respect to the nearby reference star—turbulence could make the two stars jump around on the sky, but as long as they jumped around in tandem, the faint signature of a planetary companion should still be detectable. It would be a neat trick if it worked.

A new acronym was coined to replace the ill-fated TOPS: henceforth we had ASEPS, which stands for Astronomical Study of Extrasolar Planetary Systems. It was no TOPS, but it would have to do. The only thing worse for a NASA program than having no budget is having no acronym.

Once again we were told that NASA was about to have an open competition to choose a design for the orbiting planet-search telescope. However, the PSSWG had learned the hard way that NASA's plans for selecting the final design were perpetually imminent but never executed, to our great frustration. NASA's ASEPS managers could only be even more frustrated than we were. The ASEPS program was the fourth item on the Solar System Exploration Division's wish list at a time when funds were very tight. Without the hundreds of millions of dollars needed to pay for the telescope, there was not much NASA could do, except make more plans. The race to build the orbiting planet-search telescope had been on hold for several years, waiting for its turn to start.

APRIL 22, 1994:   NASA's space-based plans might be on hold, but ground-based astronomers were continuing to advance. Alexander Wolszczan had continued to monitor his pulsar and had found the final proof that PSR1257+12 was orbited by several objects. Because the two multiple-Earth-mass planets had orbits so close together, gravitational forces between them should alter their orbits, in the same way that Neptune affected the orbit of Uranus and thereby revealed its existence. Because of the precise timing information provided by the pulsar's radio wave clock, and because the two planets fortuitously gave each other gravitational kicks at just about the same location on their orbits, these orbital changes could be seen after just a few years. The predicted orbital changes were seen by Wolszczan—there could be no further doubt about the reality of the pulsar planets.

To top it off, Wolszczan's *Science* article pointed out that with the additional years of data, he had now found evidence for a *third* orbiting body, with a mass no smaller than that of the Moon, orbiting even closer to the pulsar than the two multiple-Earth-mass bodies. Pulsar PSR1257+12 had gained a third "planet." The score now stood at Pulsars 3, Normal Stars 0. Wolszczan even hinted at the possibility of yet a *fourth* planet in the system, a Saturn-mass planet with an orbit as large as Pluto's. The Pulsars were beginning to run up the score. It was getting harder and harder to cheer for the Normal Stars.

MAY 26, 1994: Daniel Goldin gave a major address to the American Geophysical Union meeting in Baltimore, posing several intriguing "what if" questions. "What if," Goldin asked, "while protecting the precious resources on our own planet, and exploring our own solar system, we also commit to study the universe with powerful observatories to do something extraordinary? . . . What if we used them to seek and discover planets around nearby stars? Wouldn't that very knowledge of their existence transform our understanding of our place in life, of our place in the universe? Provide hope? Shouldn't we set our sights on finding out whether there is a nurturing environment beyond our solar system?"

Goldin then showed the audience a single slide—the Earth as seen by the *Apollo* astronauts, a blue and white planet hovering in the blackness of space. Goldin posed the challenge of NASA being able to return an image like that sometime in the next 25 years, not of the Earth, but of an Earth-like planet orbiting around another star.

"How do you think it might impact how every person on this planet thinks about themselves?" he asked. "Think about it. Every society leaves behind a legacy. . . . Perhaps, just perhaps, the next generation's legacy will be an image of a planet 30 light years from Earth."

Goldin raised the possibility that this extraordinarily difficult challenge could serve as a unifying theme for all of NASA's space science efforts. The latter idea was probably too much to hope for, but at the least, Goldin was extremely interested in finding extrasolar planets. Surely Goldin's NASA would soon put the TOPS/ASEPS program into high gear.

SEPTEMBER 12–13, 1994: The PSSWG pleaded with NASA to select a design for the orbiting planet-search telescope and to get this key project moving. Incredibly, NASA still had not taken this first step to get the TOPS/ASEPS program off the ground, in spite of numerous well-intentioned plans.

Roger Angel of the University of Arizona presented an idea that challenged the assumption that only a specialized space telescope would be able to detect extrasolar planets directly by taking images of the region around the target star. Angel had published a paper in *Nature* earlier in the year showing that the new generation of giant ground-based telescopes would be capable of doing the job, provided that an amazing trick was performed to minimize the blurring effects of the atmosphere that forced the original move to space. The trick was called *adaptive optics,* an idea that had been around for many years but had only recently been made to work by an Air Force laboratory. Adaptive optics relied on the use of electronically driven optical components that could remove the distortions in a bright star's image caused by the passage of the star's light through the Earth's bubbly atmosphere with the consequent bending of the star's light rays. By moving an array of mirrors quickly and continuously back and forth as needed to correct for the instantaneous distortion, the star's image could become as sharp as was theoretically possible for that size telescope, in spite of the fact that it was on the ground and not in space. The region of the sharpened stellar image would extend out to where giant planets were expected to orbit, based on our Solar System, thereby enabling extrasolar Jupiters to be detected directly around a handful of nearby stars.

Adaptive optics promised space-based performance at a ground-based price, an enticing proposition. Angel's idea was breathtaking. The PSSWG knew that if a way could be conceived to perform a challenging task from the ground as well as from space, then the ground-based effort was sure to win out, because space-based efforts always took a long time to implement and a lot more dollars.

There was a problem, though. Angel wanted to develop his adaptive optics technique on several 6.5-meter mirrors he was building at Arizona. Angel was an accomplished ground-based telescope designer with the responsibility for building a succession of progressively larger, ultimately 8-meter-diameter mirrors at his innovative mirror factory, located under the concrete seats of the University's football stadium. One 6.5-meter mirror would replace the segmented mirrors of the Smithsonian's Multiple Mirror Telescope (its acronym, MMT, would be retained—the name would be changed to the Monolithic Mirror Telescope), and another would go to Chile for the Carnegie Institution's Magellan Telescope project. Angel proposed that NASA support his plan to develop adaptive optics on these two telescopes.

The problem was that NASA had already bought into the Keck Observatory and had yet to find any money to develop specialized planet-finding instruments for Keck, much less to get involved with other ground-based telescopes. Angel had run into a stone wall.

The PSSWG returned predictably to the question of Keck interferometry and the Palomar testbed. After testing at its mountaintop site in Chile, the European Southern Observatory had just deferred its plans to link four mammoth telescopes into an interferometer, and some PSSWG members felt there might be a lesson here for Keck—the testbed needed to be moved to Mauna Kea as soon as possible, to find out if inteferometry could be made to work for the Kecks for a reasonable amount of money.

NASA's managers must have been as tired of hearing the same old refrain from the PSSWG as we were of singing it at every meeting. Time was running out, and nothing seemed to be happening.

No one knew this more than Charles Elachi, head of the Space and Earth Science Directorate at JPL, who was responsible for keeping a full pipeline of science projects for JPL. Elachi had been involved in the early phases of defining the TOPS program, encouraging us to "think big" and not to be shy about asking for whatever we felt was necessary to detect extrasolar planets. Elachi decided it was time to give the search for extrasolar planets a big shove forward.

12

THE ROAD GOES EVER ON AND ON

The ancient covenant is in pieces; man knows at last that he is alone in the universe's unfeeling immensity, out of which he emerged only by chance.

—Jacques Monod (1910–1977)

JANUARY 19, 1995: A "Dear Colleague" letter from NASA Headquarters arrived. To simplify mass mailings, NASA uses letters addressed to the all-purpose name "Colleague" to pass information to the science community. These letters, needless to say, have their own acronym—the DCL. A DCL usually meant that something important was happening; NASA's science managers are much too busy to write letters without a good reason.

The DCL advised that, not surprisingly, the long-planned selection of a design for the orbiting planet-search telescope had been postponed once again. But that was not all. The Solar System Exploration Division had asked JPL to undertake a six-month study of the options for finding new solar systems and to propose a "road map" for how to go about the adventure. The study would be chaired by JPL's Charles Elachi and would involve competitively selected teams of scientists. The outcome of the road map exercise would determine the fate of the orbiting telescope and of the rest of the TOPS program.

The TOPS plan, less than three years old, was about to be updated, but this time Elachi would be in charge. By the time Elachi was through, *road map* would be the newest buzz word at NASA Headquarters.

JANUARY 20, 1995: Prompted by a question from David Black, I published a paper in *Science* that predicted where to find extrasolar Jupiters around nearby stars. Black knew about the success of my solar nebula models at "predicting" the well-known location of Jupiter in our Solar System, and he wanted to know if that location would change for a lower-mass star. Most of the stars in the Sun's neighborhood are considerably smaller in mass than the Sun, by factors of 2 or more, and since these stars were the likely targets for astrometric planet searches, Black's question was an important one to answer.

The conventional wisdom was that because a lower-mass star is not as hot or as bright as the Sun, ice particles could orbit much closer to such stars without being in danger of being heated enough to turn into gas molecules. Because solid ice planetesimals were thought to be a key ingredient for making giant planets, it could then be expected that extrasolar Jupiters would be found in smaller orbits about low-mass stars than Jupiter's orbit in our Solar System. Such a prediction had profound implications for the optimal technique to be used to find extrasolar planets: By virtue of their mass, extrasolar Jupiters should be found first, and different search methods were sensitive to different orbital distances. If the extrasolar Jupiters had orbits that were close to their stars, then the spectroscopic method would be best suited for their detection. This is because smaller orbital separations paradoxically result in larger orbital velocities and hence in larger, more easily detectable Doppler shifts. However, if extrasolar giant planets should occur at large distances more like that of Jupiter, then the astrometric method would be better suited, because a more distant orbit means the planet is sitting farther out on the teeter-totter and can make the star wobble back and forth a greater distance.

However, the conventional wisdom was based on a faulty assumption—that planet formation proceeds at a temperature determined solely by the brightness of the star that is the final result of the star and planet formation process. Astronomical observations of suspected planet-forming disks had shown that the likely environment for planet formation was not a vacuum permeated with planetesimals being heated by their star's light, but rather was a dense, dusty protoplanetary disk, with heating and cooling processes all of its own. Observations could reveal the surface temperatures of these disks but could not penetrate deep inside the disks to reveal how hot it was where the planetesimals orbited. That is where theoretical models came into play.

When I lowered the mass of the central star in my computer models of protoplanetary disks, it turned out that the disk temperatures decreased somewhat, but nowhere near as much as would have been expected on the basis of the conventional wisdom. The reason was that

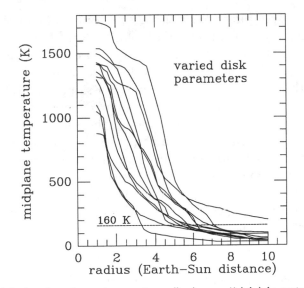

FIGURE 16. Calculated maximum temperatures (in degrees Kelvin) in protoplanetary disks with a number of different assumptions about the central star's mass, the disk's mass, and other properties. Dashed line is the temperature at which ice condenses or sublimates, 160 degrees Kelvin, which equals –113 degrees Centigrade. These protoplanetary disks are too hot to allow ice to remain in solid form, ready for planetesimal formation, except outside several times the Earth-Sun distance (1 astronomical unit = 1 AU). (Adapted from Alan Boss, 1996, *Astrophysical Journal*, volume 469, page 906.)

the disks largely determine their own temperature through processes occurring within, while the star's radiation was largely blocked from the disk's interior by the presence of gobs of dust. According to my models, even for a star 10 times less massive than the Sun, the ice condensation point should be located only about 25 percent closer to the star. It thus appeared that low-mass stars should have giant planets with orbits only a fraction smaller than Jupiter's.

An unstated assumption was that planets would be found orbiting at the location where they formed. Ignoring Goldreich and Tremaine's estimate of frighteningly rapid orbital migration of Jupiter in the presence of the solar nebula, Jupiter's formation at its present location could be explained by my models, and Wetherill had had no trouble making Earths and Venuses in the right place. The Solar System seemed to be happy to have been created more or less in place, and it seemed unsporting to assume otherwise for as yet unseen planetary systems.

The next day, January 21, Geoff Marcy kindly sent me a congratulatory e-mail message. While the *Science* prediction made his job of finding extrasolar Jupiters by spectroscopy look to be even harder to accomplish, Marcy pointed out that he had refined his methods at the Lick Observatory to the point that he expected to be able to see extrasolar Jupiters orbiting out at Jupiter's distance with data taken in the next two or three years. He was following a sample of 100 stars and could measure their velocities to within a few meters per second, about the pace of a fast walker. Jupiter's pull made the Sun move around at 13 meters per second, so Marcy would easily be able to find new Jupiters, if they were there as I had said. After I read Marcy's e-mail, my main worry was that he might not find anything at all—after all, his 1989 search for brown dwarfs had come up empty handed. Marcy seemed to have a knack for proving companions weren't there. I fervently wished him better luck this time.

JANUARY 23, 1995:   Daniel Goldin initiated a seminar series at NASA Headquarters intended to help define a suitable long-range goal for his agency. Goldin was in the midst of such a thorough shake-up of NASA that he was questioning what its reasons for existence were and whether these reasons would be sufficient to carry the agency well into the next millenium.

After the glory years of the *Apollo* program, which was at once a scientific and a human exploration crusade, to many scientists NASA seemed to lose its way. The *Apollo* program itself was scrubbed in 1972 after six lunar landings, leaving several 363-feet-tall *Saturn V* rockets without a future other than as museum exhibits. The quest to develop a reusable launch vehicle that would provide frequent, economical access to outer space metamorphosed into the Space Transportation System, or Space Shuttle. While the Shuttle has accomplished much, at a cost of 10 thousand dollars to place a pound of equipment or human being in orbit, it did not achieve its original promise of providing economical access to space. The next major engineering challenge for NASA became the 27 billion dollar International Space Station, a project fraught with budgetary problems, multiple redesigns, and the difficulties of building the most expensive object ever assembled—all with an international team of partners. President George Bush's ambitious plan for putting human beings on Mars seemed like a return to a Kennedy-like vision for NASA, but the President's plan was quickly dropped once Congress looked at the price tag. The perceived need to reduce the federal budget deficit in the 1990s brought everything that NASA and the rest of the government did under severe scrutiny.

In this murky and uncertain atmosphere, Goldin sought a new clarity of vision for NASA's future, and he turned to some of the world's leading scientists for help. For the first seminar, Goldin invited the University of Massachusetts's Lynn Margulis and the Salk Institute's Leslie Orgel to speak about primitive life forms and how we would recognize them. The seminar was held in the high-tech lecture hall of NASA's new building on E Street Southwest, a few blocks south of the Mall and Capitol Hill. A professional crew videotaped the seminar, giving the audience the sensation of sitting in a Hollywood studio during the taping of a network talk show. Goldin and France Cordova, NASA's chief scientist, moderated the event, which was shown to the field centers by the NASA television channel.

Goldin was intrigued by Margolis's and Orgel's presentations. The next month's seminar was entitled "Living Places in Other Solar Systems," with Anneila Sargent and NASA Ames's Christopher McKay as speakers. It seemed likely that the NASA Administrator was going to be a strong supporter of science and that the search for other Earths would be high on his list of priorities for NASA.

FEBRUARY 9, 1995:   Benjamin Zuckerman and his colleagues published a letter in *Nature* suggesting that extrasolar Jupiters may be rare. Zuckerman had searched for radio wave emissions from carbon monoxide gas molecules that might be orbiting around 20 nearby stars that he believed had ages in the range of 1 to 10 million years. If these supposed young stars had giant-planet-forming disks of hydrogen and helium, they should also contain enough carbon monoxide gas to be detected. About half the stars showed evidence of having carbon monoxide, but the weak intensity of the emission implied that the total amount of gas in orbit about these stars was nowhere near enough to form the giant planets in our Solar System. Zuckerman inferred that if these stars had managed to rid themselves of the gas in their disks by the time they were a few million years old, then there may not be enough gas left to make Jupiters. By the time that the ponderous giant planet formation process produced the 10 Earth-mass cores needed for pulling in the disk gas, the gas was already gone. Looking for Jupiter-sized planets, guided by the example of our Solar System's peculiar team roster, might then be a cruel hoax perpetrated by Mother Nature.

Zuckerman's letter was accompanied in *Nature* by a commentary written by George Wetherill. Wetherill acknowledged that theorists had yet to come up with a surefire, fast way to make Jupiter. Observations of disks around certified young stars by the University of Massachusetts's Stephen Strom and his colleagues had shown that these disks had lifetimes that were in some cases as short as 100,000 years, upping the ante considerably for having to make giant planets quickly. Al Cameron had dropped the idea of rapid formation through gravitational instability of a gas disk in the late 1980s, soon after Stony Brook's Jack Lissauer suggested that Mizuno's 10 Earth-mass seed cores could be built within a million years. Lissauer's suggestion required that the disk have a considerable excess of icy planetesimals at Jupiter's distance. But such a heavyweight disk would also create heavyweight planets in the terrestrial planet zone, so Lissauer advocated a nonuniform distribution of matter in the disk to solve that problem. Wetherill argued that appealing to such a distribution was ad hoc and therefore unappealing, and that for all we knew, Jupiters really were rare. Only observations could decide the issue, and Zuckerman's results were discouraging.

APRIL 11, 1995:   The NASA Keck Review Team spent several morning hours on the summit of Mauna Kea touring the IRTF and the Keck Observatory, where the second Keck telescope was rapidly taking shape. The second dome, eight stories high, was completed, and the superstructure for the second telescope was largely built.

That afternoon we heard that all plans for Keck extras, including specialized planet-finding instruments and interferometry, were on hold, pending the outcome of Charles Elachi's road map exercise. Regardless, the second Keck would be ready in less than two years, and it was therefore time to plan for its initial use by NASA's planet finders; at a cost of around one dollar per second, Keck time was an extremely valuable commodity that could not be wasted. As was the general practice in astronomy, a NASA committee would have to be formed to review proposals for using the Keck telescopes and to assign the Keck observing time. Peer review would rule, with searching for extrasolar planets being the highest-priority observations, followed by studies of protoplanetary disks and, finally, by traditional planetary astronomy.

The next day Mike Shao told us that JPL's testbed interferometer, in spite of the PSSWG's advice, was now operational on Palomar Mountain and was being used to test his ideas about doing astrometry. Shao planned to be able to measure astrometric wobbles at Palomar about a factor of 3 times smaller than those that George Gatewood could detect at Pittsburgh. With several outrigger telescopes on Mauna Kea, Shao hoped to do another factor of 3 better. Those two jumps would get him into the range of being able to find extrasolar Neptunes. The Kecks themselves would not even be needed for the astrometry but would be necessary if the interferometer was to be used to take pictures of entire planetary systems, rather than just to search for the wobbles produced by the biggest planets.

Shao's plan was to continue using the low-altitude Palomar testbed to prove the basic interferometric techniques, and then, provided that tests for the seismic stability of Mauna Kea were successful, the outrigger interferometer would be ready for building on Mauna Kea.

APRIL 18–19, 1995: Charles Elachi's road map process had spawned yet another acronym—the Exploration of Neighboring Planetary Systems, or ExNPS (pronounced "ex-nips"). Elachi held a workshop at JPL in Pasadena for the three teams of scientists who were preparing their own plans for finding planets.

A competition had been held to determine the composition of two of the three teams; by fiat, JPL had control of the third team, which was led by Charles Beichman. Six proposals to be a part of the ExNPS study were received and reviewed by a group of JPL engineers and PSSWG members; the two winners were proposals for large teams headed by Roger Angel and Robert Reasenberg.

The workshop was intended to kick off the three studies by providing a common knowledge base. The two days were spent listening to tutorials about the planet formation process and about the potpourri of techniques that could be used to detect extrasolar planets.

Elachi had high hopes for the ExNPS road map and within two years intended to be spending at least 10 million dollars a year at JPL to develop planet-finding telescopes. Elachi's time line called for the study teams to finish their work in two months, at which time an integration group would take a month to assemble the three studies into a single, coherent road map. This was fast-track planning indeed, compared to the years it had taken to draft and complete the TOPS report.

APRIL 20, 1995: Roger Angel and his colleague Adam Burrows published a commentary in *Nature* on my prediction about the expected orbital distances of giant planets. They pointed out that the result boded well not only for astrometric searches, but also for the direct imaging search that Angel hoped to initiate through his adaptive optics program for the 6.5-meter telescopes—the farther away the extrasolar Jupiters were from their host stars, the easier it would be to separate their light from that of the star and thus to find them.

Angel also raised anew the idea first proposed in 1978 by Bracewell to use a space infrared interferometer to search for extrasolar Jupiters. In back-to-back papers in *Nature* in 1986, Bernard Burke and Angel and his colleagues had realized that by updating Bracewell's design, it was conceivable that either a large space interferometer or a 16-meter space telescope would be able not only to image Earths in orbit around nearby stars, but also to study the planets' *spectra*. That is, these powerful telescopes would be able to look in detail at the wavelength dependence of the light given off by the extrasolar planets themselves, which would tell us about the composition of the planets' atmospheres. Water and ozone, for example, have a telltale effect on light emitted by a planet's atmosphere at infrared wavelengths. If the atmosphere of an extrasolar planet contained water and oxygen, there was a good chance that it supported some form of life. Angel threw out the challenge of building such a space infrared telescope as a fitting endeavor for NASA.

MAY 18, 1995: Peter van de Kamp, who had retired in his native Holland, died peacefully at the age of 93.

Van de Kamp passed away at a time when the only proven objects of planetary mass were in orbit around pulsars. He still believed that Barnard's star had planets, even if no one else had bothered to invest several decades of time searching like he had. Long-term studies published in 1995 by Philip Ianna of McCormick Observatory and George Gatewood of the Allegheny Observatory still found no evidence of a periodic wobble in Barnard's star. In fact, Barnard's star was being used by the *Hubble Space Telescope* Astrometry Team as a reference star because of its predictability—if Barnard's star suddenly jumped, the Team had a good clue that something must have gone wrong with *HST*.

One of the last things van de Kamp had said to Gatewood was, "George, you should spend less time worrying about systematic errors, and more time making them." Van de Kamp felt Gatewood should get on with the business of making astrometric measurements, and not spend so much time refining his increasingly precise techniques for measuring stellar wobbles. In actuality, Gatewood was doing both—maintaining an ongoing astrometric survey of 20 nearby stars and planning for even higher precision astrometric searches on Mauna Kea and in space.

JUNE 15, 1995: The American Astronomical Society was meeting in Pittsburgh. At a final-day session composed of papers submitted too late to make the meeting deadline, UC Berkeley's Gibor Basri announced that he had found the first evidence for a brown dwarf star that passed the lithium test. Basri and his colleagues Geoff Marcy and James Graham had done the needed follow-up observations of John Stauffer's brown dwarf candidate, Palomar Pleiades 15, and had struck paydirt. Spectra of PPl 15 taken with the HIRES spectrometer on the Keck telescope had revealed the unmistakable presence of lithium in PPl 15's atmosphere.

PPl 15 had passed the lithium test. PPl 15 must be a genuine brown dwarf, though its mass was still uncertain. Ten long years after the George Mason University conference on brown dwarfs, we had a real one at last.

JUNE 26–27, 1995:   The ExNPS Integration Team met at Beichman's Infrared Processing and Analysis Center on the Caltech campus. The time had come to synthesize the results of the three ExNPS teams into a single road map.

I was expecting a bloodbath, with all the vitriolic arguments of the PSSWG years compressed into a single, frantic, two-day meeting. Elachi had warned us at a telephone conference a few weeks before that the road map could not be an all-inclusive hodgepodge that satisfied everyone—real decisions would have to be made, and a specific, directed plan had to be decided on.

Elachi began the meeting by pointing out that Wesley Huntress would have to balance the needs of the ExNPS planet-search effort with those of two other groups: the HST and Beyond (HST&B) Committee, and the Space Interferometry Science Working Group (SISWG). The former was chaired by my Carnegie Institution colleague, Alan Dressler, and the latter by SUNY Stony Brook's Deane Peterson. Dressler's group had been charged with coming up with a concept for a successor to the Hubble Space Telescope, which was planned to be shut down in 2005, just a decade away. Peterson's SISWG had been wrestling with the issue of the two competitors for the AIM mission, Shao's OSI and Reasenberg's POINTS.

Huntress had said in effect that there was no way that NASA could afford to fly everything these three groups wanted, but he would try to combine similar concepts and missions wherever possible in order to get more bang for the buck. Both Dressler and Peterson were present at the Integration Team meeting to aid in the process of identifying elements common to ExNPS, HST&B, and AIM.

Angel led off with the presentation of his team's results. The center-piece was Bracewell's space interferometer, though modified to do a better job of canceling out (nulling) the light from the central star, in order to make the planets stand out even better. The interferometer would function at infrared wavelengths, around 10 microns, for two reasons: first, warm planets are only about a factor of a million times fainter than their stars at infrared wavelengths, rather than the factor of a billion times fainter at visible wavelengths; and second, several key atmospheric molecules are more easily detected in the infrared than in the visible—notably, carbon dioxide, water, and ozone. Angel suggested that the inteferometer consist of four 1-meter-diameter telescopes mounted on a 50-meter-long beam, and he estimated the cost at less than 2 billion dollars. To eliminate unwanted infrared emission from the Solar System's dust, the interferometer would have to be launched on a looping trajectory to Jupiter's orbital distance, outside the asteroid belt, where interplanetary dust is much less prevelant. With these parameters, the interferometer would be able to image planetary systems like our own around the several dozen Sun-like stars that are the closest to us.

Before such an ambitious telescope could be flown, however, NASA would first have to test some of the critical parts in space, such as the stability of the 50-meter-long beam. There would also be a need to learn whether or not nearby stars typically have as little interplanetary dust in orbit as the Solar System does (the zodiacal light), or as much as Beta Pictoris does. The latter would be a potential showstopper, because warm dust radiates at infrared wavelengths, and a Beta Pictoris dust disk would largely blind an infrared search for planets embedded within the disk.

Beichman's presentation had a lot of overlap with Angel's, in particular with respect to the large infrared space interferometer. This meant that at least two out of the three teams agreed on the main goal of the road map, and that was all Elachi needed. The road map's destination was fixed without any blood being shed.

As a part of Beichman's team, David Tytler of UCSD presented the case for gravitational microlensing, which he predicted would be used to claim detections of extrasolar planets in just another year or two. Microlensing involves detecting faint stars or planets indirectly by the effect their gravity has on bending the light coming from a distant star. If a faint foreground star passed between the Earth and a distant star, there would be a roughly month-long brightening of the distant star. If, in addition, the foreground star was orbited by a planet at the right distance, the planet would cause an additional brightening of the background star's light, but only for a few days or less. Remarkably, even Earth-mass planets would cause a detectable brightening. Several groups were already busily hunting for microlensing events.

Reasenberg was the third to present his team's plan, but the majority had already voted. Furthermore, the Astrophysics Division at NASA had recently downselected Reasenberg's POINTS concept from consideration for the AIM mission—now only Shao's OSI design survived. That also meant JPL had won over the Harvard-Smithsonian Center for Astrophysics for the jobs that would go along with AIM. Reasenberg was understandably glum.

With the destination fixed, it was left to fill in the sidetrips along the way. Elachi was happy to achieve the consensus support of the Integration Team by including a long list of excursions. The ongoing spectroscopic (Cochran, McMillan, and Marcy) and astrometric (Gatewood) searches must be continued, of course, and by late 1996 the 10-meter Keck telescopes would be available to NASA. Other ground-based methods with exceptional promise, such as Angel's adaptive optics and gravitational microlensing, were included as important elements of the plan.

When all the possibilities for detecting planets from the ground had been exhausted, the search would move to outer space. Several interferometric space telescopes would be built, the first intended to test the basic concept of combining the light from several small telescopes strung along a 10- to 20-meter boom. Only then would the ultimate interferometer be flown in space, a gargantuan 100 meters in length and capable of producing images of Earth-like planets within a dozen hours after it was first turned on. By Elachi's reckoning, the infrared space interferometer should cost about a billion dollars—pricey, but an absolute bargain when compared to the 10 billion dollar cost of a new aircraft carrier.

JULY 17–18, 1995: The PSSWG met at NASA Headquarters for what would turn out to be its final meeting. The primary purpose was to review Elachi's road map. Members of the Integration Team presented the multifaceted road map, which had become even more refined in the weeks since the Integration Team meeting. Not everyone was in agreement about the road map, however, even on the Integration Team itself.

Reasenberg presented a "minority report" about the ExNPS road map, calling into question the ambitious goal of the space infrared interferometer. Reasenberg's uneasiness centered on the danger of asking for an extrasolar planet-characterization mission before any planets had been proven to exist around nearby stars, on the uncertain total cost (one billion dollars—or five billion?), and on the need to fling the delicate interferometer out to Jupiter's orbit so it could do its job. As an alternative, he proposed that the AIM mission be restricted to searching for planets and be flown as soon as possible; the space infrared interferometer would be put on hold, pending the outcome of AIM's astrometric planet search.

Angel responded that he believed Goldin was serious about being willing to spend a billion dollars to do something really spectacular—to not only find Earth-mass planets, but to learn something about their atmospheres and the possibility of life on them. He also pointed out that European astronomers had independently derived a similar concept for an infrared space interferometer, raising the competitive stakes considerably. While NASA was willing to share with the Europeans the costs of flying expensive space missions, in the end the one who pays most of the bill usually gets most of the credit (as the Europeans found out to their chagrin after the spectacular successes of the NASA-dominated *IRAS* and *HST* telescopes). Cooperation was fine, but NASA wanted to make sure that it was the leading player whenever possible, minimizing the times that it had to play second fiddle to the European Space Agency.

In the same vein, Beichman worried that NASA must "own" the first extrasolar planets found—not only were the Europeans a threat in this regard, but a discovery by an American search program that had not been supported directly by NASA might be viewed as a critical failure of the new NASA initiative. The whole ExNPS program might crash if the Europeans, or UC/Caltech astronomers using the privately funded Keck I telescope, found the first planets. NASA just might lose the race before ExNPS even got started.

In a surprise visit the next morning, Daniel Goldin came in and talked with the PSSWG about his vision for the ExNPS road map. Goldin wanted a road map that was driven by the scientific goals of finding new planets, rather than by the technological urge to build big, expensive, one-of-a-kind items. He envisioned a program that would start small and build on its successes, making incremental improvements, guided the whole way by the scientific questions being answered. In his opinion, the TOPS report had failed to state clearly a visionary goal to which the general public could respond enthusiastically.

As the chair of the PSSWG, it fell to me to summarize the PSSWG's reaction to the ExNPS road map in a letter report to Jürgen Rahe, head of the Solar System Exploration Division and a long-time advocate of extrasolar planet detection at NASA. Rahe, a cometary scientist, had

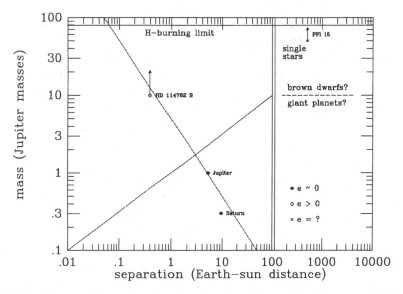

FIGURE 17. The search space for extrasolar planets and brown dwarf stars, as of mid-1995. The only brown dwarf candidates were HD 114762 B, whose mass was thought to be much greater than the lower limit shown, and the single object PPl 15, which lay perilously close to the hydrogen-burning limit of 80 Jupiter masses. There were no candidates for extrasolar planets. Diagonal lines illustrate the detection biases of searches by astrometry (dash-dot) and by Doppler spectroscopy (dashed), each of which can only detect planets lying above their respective line. The former method favors detecting planets with large separations, while the latter is biased toward planets with small separations. (Courtesy of Alan Boss.)

left his position as a professor and observatory director at the University of Erlangen-Nuremburg in Germany to join first JPL and then NASA Headquarters.

While there were some dissenters, by and large the PSSWG thought that the space infrared interferometer was an inspired concept because of the beauty of combining direct detection of Earth-mass planets with enough light-gathering power to examine the planets' atmospheres. Surely this fantastic payoff would be attractive to the American public, whose tax dollars would be required for its realization. The PSSWG also urged NASA to work to find a way to reduce the cost of the interferometer to well below a billion dollars—Goldin's dictum of NASA having to do more with the limited funds available seemed to preclude any new billion-dollar missions.

The PSSWG signed off on the ExNPS road map and soon thereafter quietly went out of business.

AUGUST 21, 1995:   The August issue of the journal *Icarus* arrived at the DTM library and contained the final results of the 12-year-long search for extrasolar planets by the Canadian group now led by Gordon Walker. After over a decade of painstaking spectroscopic observations of 21 stars, the group had come up empty handed; their observations revealed no evidence for any planets more massive than a few Jupiter masses with orbital periods less than 15 years, a bit longer than Jupiter's period of 12 years. Walker and his group gave up—they had had enough.

Walker concluded that, considering his negative results and those of Cochran, Marcy, and McMillan, Jupiter-mass planets simply were not as common as might be guessed on the basis of our Solar System. Wetherill's anthropocentric argument about why human beings could not exist in a solar system without a Jupiter seemed to be supported by Walker's observations. It seemed that theorists who had been worrying about how and where to make Jupiters were barking up the wrong tree. From now on, the search for extrasolar planetary systems would have to concentrate on the much more difficult job of finding Uranus-mass planets.

After more than 50 years of looking for planetary-mass companions to normal stars, there was absolutely nothing to show for the effort.

13

THE SWISS HIT

Many attempts are now under way to detect fully formed planets around nearby stars. . . . None of them has succeeded yet, but we are clearly on the verge of being able to detect at least Jupiter-sized planets around the nearest stars—if there are any to be found.

—Carl Sagan (1934–1996), January 20, 1994

SEPTEMBER 8, 1995: The package bearing the manuscript looked innocuous enough—just another big brown envelope sitting in my mail box at DTM. It had been hand delivered to DTM by a messenger, who drove up to northwest Washington from *Nature*'s editorial office in the National Press Building, a few blocks from the Mall and the White House. *Nature*'s astronomy editor, Leslie Sage, had called me earlier in the day to see if I was willing to take on the task of reviewing a paper claiming the detection of an extrasolar planet. I readily agreed.

The paper had been submitted to *Nature* about a week before by a pair of Swiss scientists from the Geneva Observatory: Michel Mayor, the veteran of spectroscopic searches for companions to stars, and his young colleague, Didier Queloz. The year before, Mayor and Queloz had begun using a new spectrometer at the Haute-Provence Observatory, located about 60 miles northeast of Marseille, France at an elevation of 2,000 feet. The new spectrometer was far superior to the device that Mayor and Duquennoy had used to perform their decade-long search for binary companions to all solar-type stars in the Sun's neighborhood—the new spectrometer could measure Doppler shifts for bright stars as small as 13 meters per second, a factor of 20 times better than before. That level of accuracy meant that Mayor now had available a tool that could detect Jupiter-mass planets.

Mayor and Queloz were motivated to look for extrasolar planets by the failure of the Canadian group to find *any* multiple-Jupiter-mass planets in their sample of less than two dozen stars. Mayor decided to accept the challenge of looking for extrasolar planets, but around a much larger sample of stars. The work with Duquennoy put Mayor in an enviable position—in the process of searching for binary companions, Duquennoy and Mayor had searched some 570 nearby stars of solar mass or a bit less. Mayor decided to concentrate the new search on 142 of the stars that had shown no evidence for binary companions with the old spectrometer. Mayor and Queloz thus started out with a large number of nearby stars that seemed to be single, Sun-like stars, the most promising places to look for planets, if the Solar System was any guide at all.

With the new spectrometer in place and a convincing case for exciting science, Mayor and Queloz were given a substantial block of observing time on the 1.93-meter-diameter telescope of the Haute-Provence Observatory. The design of the spectrometer also gave Mayor and Queloz an important advantage: speed. The spectrometer was calibrated by a lamp containing thorium and argon atoms, which allowed the stellar Doppler shifts to be computed as the observations were underway, rather than after the observing session was over, as was required for calibration by the iodine cells used by some of the competing American groups.

Mayor and Queloz set to work in April of 1994. They intentionally avoided monitoring any of the 120 stars that Marcy was following. Mayor and Queloz's lengthy observing list included a star called 51 Pegasi, a star similar to the Sun in its surface temperature and mass, but older and containing somewhat more metals (e.g., iron). 51 Pegasi is located about 45 light years away in the constellation Pegasus.

In September 1994, Mayor and Queloz made their first observations of 51 Pegasi, after the constellation Pegasus became visible in the night sky. By the time they had observed it just three more times, Mayor and Queloz realized that 51 Pegasi was acting as if it could not sit still. By January 1995, they had calculated a preliminary orbit for a planet that could explain 51 Pegasi's shakes. However, the constellation Pegasus was by then disappearing from nighttime view, as the Earth's orbit around the Sun caused the constellation to lie in the same general direction as the Sun. Mayor and Queloz would have to wait six months for the Earth's orbit to bring Pegasus back up into the night sky so they could confirm their prediction.

In July 1995, the constellation Pegasus rose again above the night-time horizon, and Mayor and Queloz finally had the opportunity to double check their prediction for 51 Pegasi. They followed 51 Pegasi intensively during an extraordinary observing run of eight nights' duration. Only then were Mayor and Queloz convinced that they had truly found what they were after—the first planetary-mass object in orbit around a star very much like the Sun. On August 29, 1995 they submitted their paper to *Nature*. Meanwhile, they continued to refine their measurements of 51 Pegasi and planned another eight-night observing trip for September 1995.

Mayor and Queloz found that 51 Pegasi was being pulled back and forth by up to about 60 meters per second, a whoppingly large effect that could be clearly seen with a spectrometer with their accuracy. When I first saw their discovery data, plotted on top of the curve showing the velocity changes predicted for a star being tugged by a planet, it was immediately clear that there was no danger of this being a case of looking too hard at noisy data and deluding oneself into thinking that one could see a signal—this was obviously a *detection*. A detection, yes, but of what? Maybe it was a very low-mass star, or even a brown dwarf.

51 Pegasi appeared to be circled by an object with a mass at least half as large as that of Jupiter. The object could be much larger in mass than Jupiter only if it happened to orbit 51 Pegasi in a plane that was perpendicular to the direction to Earth (i.e., if its orbital plane was viewed face-on from the Earth). However, such an orbital orientation is considered a relatively unlikely occurrence. Think of a frisbee as representing the orbit of 51 Pegasi's companion. If one tosses a large number of frisbees randomly in the air, the chances are small of seeing one exactly face-on; most will be seen more or less edge-on. Most likely, then, the companion's mass was about 60 percent of Jupiter's mass.

Furthermore, Mayor and Queloz knew that 51 Pegasi's spectrum showed evidence that 51 Pegasi was rotating at about the same rate (once a month) as other solar-type stars and that 51 Pegasi's rotation axis was most likely oriented perpendicular to the line of sight, so that both the star's equator and the probable orbit of the planet were being viewed nearly edge-on. If that was the case, then 51 Pegasi's companion could be no more massive than half Jupiter's mass. Mayor and Queloz must then have found the first extrasolar planet orbiting a Sun-like star, based on what we know about the masses of the planets in our Solar System.

When asked to referee a short, nonmathematical paper that has been submitted for publication in *Nature*, I normally try to read the

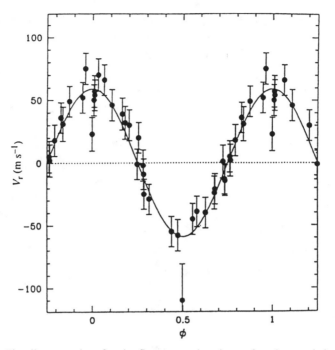

FIGURE 18. The discovery data for the first extrasolar planet, found around the Sun-like star 51 Pegasi by Michel Mayor and Didier Queloz. The solid line fit to the Doppler velocity shifts of 51 Pegasi implies the star is orbited by a planet at least half as massive as Jupiter. (Reprinted, by permission, from M. Mayor and D. Queloz, 1995, *Nature*, volume 378, page 357. Copyright 1995 by Macmillan Magazines Limited.)

manuscript and write the report in the same day, in order to complete the process in as short a period of time as possible and get the paper moving along in the publication process. The 51 Pegasi paper took a few days longer to review because it arrived in the midst of preparations for the annual meeting of the Meteoritical Society, an international convocation being held in Washington that year.

The brown envelope had arrived late on a Friday night. Saturday morning was spent with the rest of the organizing committee, getting ready for the Meteoritical Society's meeting, to be held beneath the Mall in the Smithsonian's Ripley Center. Saturday afternoon I returned to DTM and finally got a chance to read the manuscript. The fact that Mayor and Queloz had such strong evidence for a half-Jupiter-mass planet was intriguing, but the really big surprise was the unbelievably short period of 51 Pegasi's planet—a little over four days, ridiculously short compared to Jupiter's stately 12-year orbit around the Sun. The peculiarly short orbital period inferred for 51 Pegasi suggested that extreme caution was in order.

Mayor and Queloz could tell from the way that 51 Pegasi's velocity varied with time that the star and planet were moving in a nearly circular orbit about their common center of mass. Such a circular orbit would help clinch the case for the companion's being a planet rather than a brown dwarf star, were it not for the peculiarly short four-day period. Binary stars with orbital periods less than about two weeks are close enough together that their mutual gravitational forces have enough time to force the binary's orbit to become circular; by maintaining a fixed distance from each other, the two stars no longer have to undergo the constant stretching and squeezing caused by their tidal forces, like a long-married couple who have found that the best way to maintain a stable relationship is to keep a certain distance from each other. 51 Pegasi and its companion would be expected to have a circular orbit, regardless of how distorted their orbit was initially, because a billion years or so would be plenty of time to circularize a four-day orbit. The primordial eccentricity of 51 Pegasi's companion thus could not be determined and so could not be used to determine if it was really a planet or a brown dwarf.

Sunday was spent again preparing for the Meteoritical Society meeting and then attending the meeting's opening reception in the garden outside the Smithsonian Castle. Monday was a full day for me: Several of my DTM colleagues and I gave talks in a morning session, and in the afternoon there was a special session I had helped to organize, followed by a session of poster talks in the early evening. *Nature* had asked me to write up a report about the meeting, plans for which took up part of Tuesday morning, September 12, and I did not get a chance to get back to the Mayor and Queloz paper until that afternoon, a maddeningly long wait for me.

SEPTEMBER 12, 1995: I did my best to find some hidden flaw in the paper, but I could not come up with much. One other explanation for 51 Pegasi's velocity changes every 4.23 days was that maybe it was a variable star (that is, a star that pulsates in and out, as if it were breathing). Such stars are reasonably common in the galaxy, and four-day periods are typical for certain classes of variable stars, such as the Cepheids. Polaris, the north star, is a Cepheid variable with a four-day period. Perhaps Mayor and Queloz might then have been measuring the velocity of the star's surface as it periodically rose and fell, rather than the wholesale motion of the star back and forth in an orbit. The Canadian group had found such variability to be the ultimate explanation for Gamma Cephei's shivers.

Mayor and Queloz had thought of this possibility, of course, but had addressed it only sketchily in their necessarily brief manuscript. They stated that they had measured the brightness of 51 Pegasi 67 times and found that it did not vary, at least not to the precision of their measurements. If 51 Pegasi was in reality a variable star, rather than the host of an extrasolar planet, it should brighten and then dim every 4.23 days, as its radius changed from the maximum value to the minimum. As long as 51 Pegasi did not seem to be changing in brightness, as Mayor and Queloz claimed was the case, there was a good chance it was not a variable star. But maybe 51 Pegasi just so happened to be variable when they took their velocity data, but not when they monitored its brightness. Could the variable star explanation really be ruled out?

I sent my referee report by e-mail to *Nature* shortly after 6:00 P.M. and then headed down to the Carnegie Institution's Root Hall, located a few blocks north of the White House. That evening, Wes Huntress and his counterparts from the European, Japanese, and Russian space agencies would be telling the Meteoritical Society membership about their nations' plans for future planetary exploration, and as one of the evening's organizers, I did not want to miss the affair. I was bursting to tell everybody what Mayor and Queloz had found around 51 Pegasi, but I could not. A referee must maintain a strict silence about a paper until the results are released by the authors or by the journal. There was a good chance that our collective bad luck had finally changed for the better in the search for extrasolar planets, but the time was not yet ripe to make public pronouncements. That honor rightfully belonged to Mayor and Queloz, if they could prove their case.

At the same time that I was elated by the discovery, I was also disturbed by what the Swiss had found, not the least because their finding seemed to demolish the prediction I had published in *Science* earlier in the year. I had predicted that extrasolar Jupiters should be found in orbits around low-mass stars more or less similar to that of Jupiter's. Mayor and Queloz seemed to have finally found an extrasolar Jupiter, but with a 4.23-day period. This meant that 51 Pegasi's planet had to orbit at a distance 100 times closer than Jupiter does around our Sun, a startling and unexpected result. The planet would have to be located much closer to its star than already sweltering Mercury is to our Sun. It was inconceivable, based on my models and those of other astrophysicists, that a giant planet like Jupiter could form so close to a solar-type star—there was no way that any icy planetesimals could exist that close to the stellar inferno, much less at that fiery distance inside the proto-planetary disk from which planets form.

Mayor and Queloz's claim caused me some sleepless nights. Maybe it would all turn out to be a mistake, another in the long succession of failed claims for extrasolar planets. But what if the Swiss really were right after all? Then seemingly I had to be very wrong—what had I missed? Where had I gone wrong? The likely answer came to me late one night, as I lay in bed staring wide-eyed at the darkened bedroom. If 51 Pegasi has a Jupiter-mass companion orbiting very close to the star, yet my models showed that a gas giant planet could not form anywhere near that close to the star, then it could only mean one thing—some planets must migrate great distances from their place of birth. 51 Pegasi's planet must have formed more or less at its predicted location, and then been dragged kicking and screaming inward toward its star. The likely means for accomplishing this forced march was something theorists had secretly worried about for some time—interactions of giant planets with the disk of gas and dust from which they formed—but had never bothered to fold into predictions of extrasolar planet orbits, since such migrations had not been necessary to explain the present-day locations of the planets in our Solar System. The discovery of 51 Pegasi's planet might now force theorists into revealing this dirty little secret: Planets might not stay put where they were made.

SEPTEMBER 14, 1995: Rafael Rebolo and his colleagues published in *Nature* the likely discovery of another brown dwarf star in the Pleiades. Following the example set by John Stauffer with Palomar Pleiades 15, they named their object Teide Pleiades 1, or Teide 1 for short. (Teide is the name of the volcano for which the Teide Observatory is named, on the Canary Island of Tenerife.)

Rebolo and his group had found Teide 1 by the same technique that Stauffer and his group used to find PPl 15—a survey of the Pleiades star cluster at two wavelengths that would reveal the presence of any cool stars radiating primarily at near-infrared wavelengths. Teide 1, in fact, was even fainter than PPl 15, and since it most likely had about the same age, it was probably even lower in mass. Rebolo's group estimated that the mass of Teide 1 was probably in the range of 20 to 30 times the mass of Jupiter, well below the stellar cutoff at 80 Jupiter masses. Teide 1 looked good, but the acid test would be the lithium test that PPl 15 had already passed—a brown dwarf star should never become hot enough to destroy its lithium. One year later, Teide 1 passed the lithium test, too.

Teide 1 was found by Rebolo's group after they searched only a small fraction of the Pleiades. They estimated that there should be hundreds of brown dwarfs lurking in the cluster, enough to provide the good-sized sample that would be necessary for astronomers to begin to understand brown dwarf stars as a new class of stars, from a statistical point of view, rather than as isolated oddballs.

SEPTEMBER 30, 1995: Deane Peterson's SISWG finished their suddenly fast-paced deliberations regarding the AIM called for in the Bahcall Report. They had been asked by NASA's Daniel Weedman, head of the Astrophysics Division, to take no more than a few months to decide the issue. The SISWG agreed that Michael Shao's design for OSI met the requirements for AIM. In addition, the SISWG decided that OSI could serve as a testbed for some of the more ambitious technological elements required for both the ExNPS space infrared interferometer and the large infrared space telescope being advanced by Alan Dressler's HST&B committee.

A few weeks before, the Solar System Exploration Division had ceased to provide money to support the development of the various ideas proposed for a space-born planet-finder telescope (e.g., POINTS, FRESIP, ATF/CIT). The long-anticipated downselect had effectively occurred, not with the shouts of a newly selected winner, but with the sullen silence of terminated projects.

Shao had pioneered the construction of optical interferometers through the Mark I-II-III series of interferometers he had built on Los Angeles's Mount Wilson, starting in 1979. His years of efforts had paid off. A few months before, Shao had unusually rapid success in getting the Palomar testbed interferometer to make interference fringes by combining the light beams from several of the interferometer's mirrors, a key milestone signaling the ability to link the interferometer's elements into a single telescope. With the SISWG's approval of OSI as the AIM design, Shao now ruled the roost of NASA interferometry.

OCTOBER 5–6, 1995: The ExNPS Integration Team met for the final time at JPL to present the ExNPS road map to the Blue Ribbon Panel, chaired by UC Berkeley's eminent physicist, Charles Townes. The Townes Panel, composed of 20 leading scientists and policy makers, would review the ExNPS plan and make the final evaluation of the road map before it was submitted to Daniel Goldin.

Charles Elachi led off with the history of NASA's efforts to plan for the detection of extrasolar planets. It then fell to me to describe our basic understanding of the star and planet formation process. That was easily done, but the question of keeping quiet about Mayor and Queloz's 51 Pegasi detection was more of a problem. I had been telephoned earlier in the week by a British Broadcasting Corporation reporter who wanted to know what I knew about the rumors beginning to float around about a new planet detection. I had told the reporter that as a referee for the paper, I could not say anything, and I then e-mailed Leslie Sage for advice. Sage replied that it sounded like Mayor and Queloz were beginning to let the cat out of the bag and that it was up to my discretion. At JPL that morning, I had heard that Mayor would be presenting his results the next day at a meeting in Italy. So when I was asked about the rumors during the question period after my talk, I felt it was time to let the Blue Ribbon Panel know that there was a very good chance that a Jupiter-mass planet had already been discovered by the Swiss. I pointed out that the weird location of 51 Pegasi's planet implied that it must have migrated inward as a result of the sorts of interactions with the disk that astrophysicists like Doug Lin and Bill Ward had been considering for some time.

On the morning of the second day, Caltech's Shrinivas Kulkarni asked for some time to make a special presentation before the Blue Ribbon Panel went into a closed-door session to figure out how to respond to the ExNPS road map. Kulkarni was not on the Integration Team or the Blue Ribbon Panel, but he knew that an influential audience was in Pasadena, and he had a hot show to put on. Kulkarni entertained the Panel and the Team with the startling announcement that he and his colleagues had found the first brown dwarf companion to a star, not a free-flyer like PPl 15 or Teide 1, but a companion to a low-mass star called Gliese 229, located about 19 light years away from the Sun in the constellation of Lepus.

Kulkarni's Caltech and Johns Hopkins University team had found this brown dwarf companion to Gliese 229 not by spectroscopy or by astrometry, but by direct imaging—they had actually gone out and taken its picture. They were able to accomplish this feat by using a unique coronagraph that could block out most of the light from Gliese 229 and that could remove some of the blurriness caused by the Earth's atmosphere. The latter was achieved by employing the same sort of adaptive optics techniques that the Air Force had pioneered. With the

Adaptive Optics Coronagraph, developed by the Johns Hopkins scientists, mounted on the Caltech/Carnegie Institution 60-inch telescope on Palomar Mountain, Kulkarni's team began to search nearby, relatively young (billion-year-old) stars for the possible presence of brown dwarfs.

Out of the first 100 or so stars studied, one of them stood out. Gliese 229 was first imaged by Kulkarni's team on October 27 and 29, 1994, and these images revealed the presence of a faint companion. However, the faint companion could just be a distant background star or galaxy that happened to line up with Gliese 229, so the team had to wait one year and take another image. During the period October 3 to 8, 1995, the group again imaged Gliese 229 and held their breath to see if it was still in the same location—and it was. In the intervening year, Gliese 229's movement across the sky (its proper motion) carried it away from its location on the sky in October 1994, so if the faint companion had moved along with Gliese 229's companion during this period, then the companion most likely was linked to Gliese 229 and was not some random interloper. Kulkarni was so excited by the discovery that he came to JPL to tell us about it while the rest of his team was still observing on nearby Palomar Mountain.

Following the standard astronomical practice, the primary star Gliese 229 is called Gliese 229 A, while its faint companion is termed Gl 229 B. Kulkarni's group estimated that Gliese 229 B was giving off no more than one tenth of the light emitted by the lowest-mass normal star known, implying that it must be a brown dwarf. The angular separation of Gliese 229 B from Gliese 229 A implied a linear separation about 44 times greater than that of the Earth from the Sun, a bit larger than Pluto's average distance from the Sun. Because the orbital period of an object at that distance is many hundreds of years, Gliese 229 B did not appear to move appreciably along its presumed orbit in the year-long interval between the discovery images taken with the coronagraph—it may take decades of persistent imaging to begin to watch Gl 229 B's orbit unfold. But from the measured luminosity of the companion and its assumed age of about a billion years, the team could estimate that Gliese 229 B must have a mass in the range of 20 to 50 Jupiter masses, comfortably below the minimum needed to burn hydrogen and be a normal star.

But that was not all. The month before, on September 14 and 15, the team had used a spectrograph on the 200-inch Hale Telescope to collect light from Gliese 229 B. They gathered enough light to produce a reasonably good spectrum of the putative brown dwarf. The spectrum was unlike that of Stauffer's PPl 15, which looks much like that of a low-mass, normal star except for the presence of the lithium absorp-

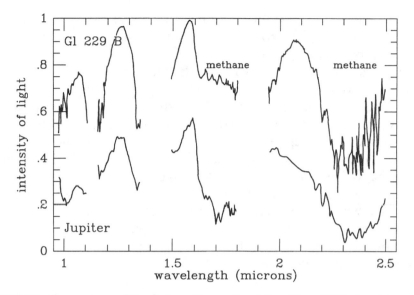

FIGURE 19. The spectrum of the first cool brown dwarf star, Gliese 229 B, found by a team of scientists from Caltech and the Johns Hopkins University. Gliese 229 B exhibits the same spectral features as Jupiter, caused by methane gas molecules. Methane cannot exist in the hotter atmosphere of a normal star. (Adapted from B. R. Oppenheimer et al., 1995, *Science*, volume 270, page 1479.)

tion lines, which reveal PPl 15 to be a genuine brown dwarf star. Instead, Gliese 229 B had a spectrum that looked not like a star, but like *Jupiter*. Gliese 229 B's spectrum showed the clear presence of methane absorption bands at near-infrared wavelengths that look just like those in Jupiter's spectrum. Kulkarni estimated that Gliese 229 B's surface temperature was about 600 degrees Centigrade, thousands of degrees cooler than the surfaces of the coolest normal stars. Gliese 229 B was considerably hotter than Jupiter, which has a surface temperature of about −100 degrees Centigrade, but Gliese 229 B was still cool enough to allow methane molecules to exist in its atmosphere.

Kulkarni's team had found the first cool brown dwarf star, Gliese 229 B, an object with properties intermediate between the hot brown dwarf PPl 15 and Jupiter.

The spectral similarity of Gliese 229 B to Jupiter emphasized how the line between stars and planets begins to blur in the regime of gas giant planets and brown dwarf stars. In fact, theoretical models of the internal structure of gas giant planets and brown dwarf stars are remarkably similar, with the main difference being the rock and ice cores indirectly inferred to exist for the former on the basis of spacecraft trajectories during NASA's robotic flybys of Jupiter and Saturn, while brown dwarf stars are modestly assumed to be completely mixed and not to contain solid cores. It will be difficult to prove or disprove the latter assumption, but, in general, a multiple-Jupiter-mass ball of hydrogen and helium gas attains a structure that is relatively independent of the way in which it was formed, as a star or as a planet. The implications for whatever else might be in the neighborhood, however, are profound: If the object is truly a gas giant planet, then there is a good chance that Uranus-like and Earth-like planets are also loitering nearby. For this reason, if no other, it is crucial to differentiate between giant planets and brown dwarf stars—the former are the guideposts to extrasolar Earths.

October 6 was a red letter day for astronomy, not only in Pasadena, but half a world away in Italy.

14

THE WORLD FINDS OUT

If you want to succeed in this world, you don't have to be much cleverer than other people, you just have to be one day earlier.

—Leo Szilard (1898–1964)

OCTOBER 6, 1995: The ninth conference in a series dealing with the astronomy of relatively cool stars was being held in Florence, Italy. Mayor and Queloz's manuscript about 51 Pegasi was still hovering in the limbo between submission and either acceptance or rejection. *Nature*'s Sage had warned them that the paper's referees had raised serious issues, but Mayor and Queloz could not wait any longer. The Florence conference was the perfect opportunity to present their dramatic results to a largely unsuspecting audience of their peers. Mayor and Queloz went ahead in Florence and revealed their wonderful secret that 51 Pegasi seemed to be circled by a half-Jupiter-mass planet.

After Mayor gave the talk, he and Queloz listened to the thunderous applause coming from the audience. Scientific audiences are normally very polite and will give a brief round of applause for even the most dreadfully boring or obscure talk, so a speaker must learn to subtract this minimal amount of applause in order to discern whether the audience is reacting with true enthusiasm. Mayor and Queloz could tell that the conference's scientists were pleased by their discovery.

Mayor and Queloz did not speak with the Italian newspaper reporters in attendence and restricted their remarks to their conference presentation because of the uncertain status of their manuscript at *Nature*. The *Washington Post* responded two days later by publishing a story buried deep inside, on the 36th page, with the headline "Italian Astronomers Say a Jupiter-like Planet Circles a Star in Pegasus," unintentionally moving the Geneva Observatory through the Mont Blanc tunnel and into Italy. The tone of the article was suitably cautious, noting that other astronomers had raised questions about the Swiss discovery.

Before the discovery of 51 Pegasi's Jupiter-like planet, there was no firm scientific proof that planets similar to those in our Solar System existed anywhere else in the universe. Many trails of evidence pointed to the likelihood of their existence, but by mid-1995 the repeated failure to find extrasolar Jupiters was beginning to erode severely the serene confidence that other Jupiters must exist—the Canadian group's null result had just been published in August. Two months later, Mayor and Queloz stunned the world with the announcement of their discovery. The cliche about the situation being darkest just before the dawn came true.

Mayor and Queloz, who were not generally known to be in the race to find extrasolar planets, beat all of the American teams to the first prize. Several of the American teams had been hard at work since the 1980s, whereas the Swiss had searched for only a year and a half, but the race went to Mayor and Queloz nonetheless.

The Swiss search strategy was influenced by Mayor's extensive background in looking for binary star companions. Binary stars are known to exist with a wide range of orbital periods, from tens of millions of years for the widest visual binary pairs to a few hours for the closest stars. Mayor was thus mentally prepared to search his data from 51 Pegasi for short period orbits, such as the four-day period found for 51 Pegasi B. The Americans, on the other hand, were looking for Jupiter clones and expected to find companions with periods of a decade or so; with that sort of expectation, in fact, you might as well not even bother to analyze your data until you have accumulated several years worth. My *Science* paper had only helped to fuel this attitude about the likelihood of long period orbits for gas giant planets. The Swiss' absence of such preconceptions about the orbital periods of companions helped enable their discovery of 51 Pegasi B.

By chance, the October 1995 issue of *Scientific American* featured an illustration on its cover showing the view from the surface of a hypothetical planet in orbit around a star in a triple-star system. Inhabitants of such a world might periodically see three suns instead of one, I had imagined in an accompanying article about the search for binary star companions to very young stars. The *Scientific American* art department decided to pick up on the theme of planets that might also exist in binary- and multiple-star systems and created the cover painting. The timing of the issue was completely coincidental but impeccable—it had just been released when Mayor and Queloz made their announcement in Florence.

While we on the PSSWG and TOPSSWG committees had been seemingly frittering away our time, debating the designs of imaginary space telescopes capable of finding extrasolar planets, arguing over the location of the Palomar testbed, and generally gnashing our teeth over the glacially slow process of pushing for the funds needed to launch a serious search program, Mayor and Queloz had quietly gone about their business and, using a modest ground-based telescope, found the first extrasolar planet.

If not for a tragic accident, the discovery of 51 Pegasi B probably would have been made by Michel Mayor and another colleague, Antoine Duquennoy. Duquennoy and Mayor had labored for years to perform the initial survey work that eventually led to the 51 Pegasi B detection, a decade-long search for stellar companions to every Sun-like star within 72 light years of the Earth. A few years after finishing this monumental task in 1991, Duquennoy died in an automobile accident on a twisting Alpine highway while trying to drive back home after a strenuous day of mountaineering. Duquennoy died on June 19, 1994, two months after the planet search began with the new spectrometer at the Haute-Provence Observatory. If not for this terrible loss, Didier Queloz might not have had the same chance to be a part of what promised to be one of the major scientific discoveries of an extraordinarily productive century.

With Mayor and Queloz's epochal discovery finally made public, the next step was to find out if anybody else could duplicate the Swiss observations and confirm that 51 Pegasi really had a planetary companion. If the discovery could not be confirmed, the Swiss claim might have to be tossed onto the heaping pile of previous failed dreams.

The race to confirm 51 Pegasi B would go to whoever had a high-precision spectrometer and could get time on a suitable telescope. Normally, astronomers propose to use telescopes through an orderly process, submitting proposals every six months for observations they hope to perform in the next year. In the case of 51 Pegasi B, there was no time to wait for the next round of telescope proposals—you just had to bump off whoever was scheduled to use the telescope and get on with the time-critical observations.

William Cochran had a spectrometer at the McDonald Observatory that could do the job, but he was attending a planetary science meeting in Hawaii during the week after Mayor and Queloz's announcement (coincidentally being held in the same plush Mauna Lani hotel where the

TOPSSWG had worked on the TOPS report). By the time Cochran could return from the Hawaii meeting, it would be too late.

OCTOBER 19, 1995: Geoff Marcy and Paul Butler managed to be the first to confirm Mayor and Queloz's extrasolar Jupiter, and they did it in record time. This time the *Washington Post* reported the story of 51 Pegasi and its planet on the front page.

Marcy and Butler had been able to get telescope time the week immediately after Mayor and Queloz made their announcement in Florence. During four consecutive nights on the 120-inch telescope at the Lick Observatory, they used their iodine cell spectrometer to follow 51 Pegasi's every tremble. They took a few days to analyze their 27 sets of data and then went public: Within two weeks of Mayor and Queloz's announcement, Marcy and Butler had confirmed the existence of 51 Pegasi B. For the first time, two independent teams of astronomers had produced robust evidence that a planet orbited around a star much like our Sun.

That night Marcy and several other astronomers were interviewed by Ted Koppel on the ABC News *Nightline* television program. Koppel was enthusiastic about the discovery of 51 Pegasi B by the Swiss and by the rapid confirmation by Marcy and Butler, and he immediately began to draw the conclusion that the discovery raised the chances that intelligent life must exist elsewhere in the universe. Koppel mentioned that scientists were already listening, hoping to find evidence of extraterrestrial intelligence, but Marcy pointed out that Congress had terminated NASA's support for Frank Drake's SETI program. Koppel joked that perhaps the same funding problems beset alien astronomers who were hoping to hear from us.

The real surprise of the *Nightline* program was the announcement made by Harvard-Smithsonian's David Latham, who was being broadcast live from the Oakridge Observatory outside Boston. Latham had been invited by ABC News to participate in the show that same afternoon, and in preparation for the broadcast, Latham had decided to do a quick analysis of his own spectroscopic observations of 51 Pegasi. While Latham's spectrometer was not capable of anything close to the same degree of precision that the Swiss and Lick astronomers had achieved, he did have 55 observations of 51 Pegasi spread out over several years, made as part of a long-term program to expand the Duquennoy and Mayor survey for binary companions to an even larger sample of stars. While Latham had but a few hours to look at his data, he stated on *Nightline* that he had evidence for a *second* planet in orbit around 51 Pegasi—there seemed to be a gradual drift in the mean velocity of 51 Pegasi that might indicate the pull of another planet. Latham was suitably evasive when asked about the mass or orbital period of the second planet, saying only that in order to be seen by his equipment, it would have to be at least as massive as 10 or 20 Jupiters and must have an orbital period longer than a year. The heady possibility arose of an entire planetary system around 51 Pegasi.

Marcy went along with Latham's surprise claim for a second planet around 51 Pegasi during the show, but he must have been a bit chagrined by all the attention being paid to Latham's highly uncertain and overly massive "planet." Marcy knew that his higher-precision spectrometer would soon yield a definitive measure of whatever else might be orbiting 51 Pegasi, and he promised an answer within a month.

OCTOBER **24, 1995:**   UCSC's Douglas Lin and Peter Bodenheimer and Derek Richardson of the Canadian Institute for Theoretical Astrophysics submitted a paper to *Nature* that sought to explain the peculiarly short orbital period of 51 Pegasi B.

There was no explanation at all for how 51 Pegasi B could have formed at its present location, 20 times closer to 51 Pegasi than the Earth is to the Sun. Evidently, the planet must have formed farther away from 51 Pegasi, where water ice could be stable and icy planetesimals could form, and then somehow been moved inward to its precarious location. This presumed planetary migration, however, had a major and disappointing implication—if any Earth-like planets had also formed around 51 Pegasi, at Earth-like distances, these planets must have been eliminated long ago, as the much more massive planet 51 Pegasi B moved inward through their orbital locations. The passage of 51 Pegasi B would either violently kick any intervening Earth-mass planets outward to the frigid regions of interstellar space, or drive them into collisions with their sun, 51 Pegasi A, or with 51 Pegasi B itself. In the latter two cases, any preexisting Earth-mass planets might be swallowed whole. Such would be the fate of a relatively small planet like Earth in a planetary system containing gas giant planets on the move.

Geoff Marcy had suggested to journalists that 51 Pegasi's orbital position might have been caused by a game of cosmic billiards. If several massive planets once existed around 51 Pegasi, and if they happened to form close enough together that their mutual gravitational pulls made their orbits unstable, then all hell might break loose. These

Jupiter-mass billiard balls would undergo a chaotic evolution, careening off on highly elliptical orbits following close encounters with each other. Eventually things would settle down again, but only after one or more planets had been kicked out of the planetary system. One planet would be left behind on a much shorter period orbit, to compensate for the angular momentum of the ejected planet, and maybe that was the origin of 51 Pegasi B. It seemed hard, however, to imagine a planet being left behind so close to its star as a result of this gravitational scattering process.

On the other hand, Goldreich and Tremaine's bogeyman of making Jupiter-mass planets migrate inward or outward through gravitational pulls from the protoplanetary disk would have no scruples about forcing a wayward planet to move its orbit right on down to the surface of its sun, where it would merge with the star and disappear. Goldreich and Tremaine did not suggest any means for saving the planet at the last minute, however—in their nightmare, the planet was doomed from the beginning.

Lin and his colleagues thought of a way out, a way to stop the relentless bogeyman from feeding 51 Pegasi A its own progeny. Gravitational interactions with the disk would continue to pull the planet backward in its orbit, thereby draining the planet's orbital angular momentum and forcing it to orbit ever closer to its sun, but only up to a point. Lin and his team recognized that when the planet came very close to its sun, another effect would come into play—gravitational interactions between the planet and the central star.

The gravitational pull of the planet on the star raises a tide on the star, just as the star's gravity raises a tide on the planet. If the star and planet were not rotating and were fixed in space, the highest tides would occur along the line joining the centers of the two bodies (i.e., the highest tide would occur when the other body was directly overhead). In the real world, the star and planet will rotate and orbit about the center of mass of their system. If the star rotates faster than the planet orbits around the star, and in the same direction, the star's rotation will carry the location of the highest tide on the star *ahead* of the line pointing toward the planet.

The planet thus orbits *behind* the tidal bulge on the rotating star. As a consequence, the planet is pulled forward by the gravitational tug of the star's tidal bulge. This interaction adds angular momentum to the planet's orbit, at the expense of the star's rotation, which then slows down. As the planet moves closer to the star, it gains more and more angular momentum from the rotating star. Eventually it reaches a distance at which the planet is gaining angular momentum from the star at the same rate that it is losing angular momentum to the disk through the Goldreich and Tremaine mechanism. At that distance, the planet silently stops, locked onto a stable orbit about its sun.

Young stars are known to rotate at least 10 times faster than middle-aged stars like the sun, so Lin's scenario made perfect sense. It takes billions of years for stars to slow down to solarlike rotation rates, as they lose angular momentum along with the mass being carried off by their winds. By that time the protoplanetary disk would be long gone, and the planet would no longer have to worry about being swallowed alive.

Tidal forces depend strongly on the distance between the star and the planet, and so they would only be expected to become important when the planet came very close to the star, on the order of a few times the radius of the star. That was exactly the case with 51 Pegasi B, which orbited its star at a distance of about eight times its present radius, a radius that was about three times larger when 51 Pegasi A was a young star.

If 51 Pegasi B could not be made in situ, then, there was at least one reasonable means to force it to march down front and sit still. Of course, our own Solar System showed no evidence of having endured this humiliating experience—if it had, a habitable Earth would have been destroyed in the process, and we would not be around to worry about it. Something must have gone "right" to prevent the disk bogeyman from ruining our Solar System. We already knew that disks around young stars have a wide range of lifetimes, from a hundred thousand years to ten million years. Perhaps the Solar System avoided the fate of 51 Pegasi's planets by having been blessed with a disk that did not last long. Soon after Jupiter gathered its complement of gas, and before Saturn could grow to more than a third the mass of Jupiter, most of the gas was gone from the solar nebula, removed by flow inward onto the protosun and outward to interstellar space. The agent that performed this disk eviction is uncertain, but we may owe our very existence to the prompt and thorough job it performed.

OCTOBER 25, 1995: The NASA Keck Review Team met at NASA Headquarters, allowing the group to hear from NASA managers about the progress on the Keck Telescopes and about the recent reorganization of NASA's headquarters. The Review Team was chaired by Michael Belton of the National Optical Astronomy Observatory in Tucson, a senior planetary scientist with extensive experience at both robotic and ground-based observations.

To meet Daniel Goldin's goals for downsizing NASA's Headquarters operations, over half of the space sciences Headquarters staff would be forced out, and the lucky survivors would find their responsibilities had increased. One of the winners in the reorganization process was Edward Weiler, who had managed the highly visible and successful *HST*. Weiler was chosen to direct NASA's efforts in a new area termed Astronomical Search for Origins and Planetary Systems, or Origins for short. This nickname avoided the use of an acronym (ASOPS) that could easily be confused with the preexisting planet-search acronym, ASEPS. Origins would encompass not only the formation of stars and planets, but the origin of galaxies and of life. Weiler would now run NASA Headquarter's efforts to find extrasolar planets, though Jürgen Rahe, still head of NASA's planetary exploration program, was expected to continue to lend strong support to the extrasolar planet-search effort he had helped to bring into existence.

The second Keck telescope would be ready for use in less than a year, and the Review Team was asked to define the guidelines for the committee that would allocate NASA's share of the Keck telescope time. A carefully reasoned exchange of e-mails ensued, regarding the genesis of NASA's involvement in the Keck Observatory. Contrary to the original plan, the funds to pay for Keck II had come out of the hide of NASA's planetary exploration program. In spite of this, and in spite of the desire of many planetary scientists to use the Keck to observe the Solar System's comets, asteroids, and planets, the Review Team agreed that the top Keck priority should be looking for extrasolar planets. Now that Mayor and Queloz had found 51 Pegasi B, there was an implicit realization that NASA would have to get back into the race as soon as possible, and the Keck Observatory was almost ready to become an important part of the race.

NOVEMBER 23, 1995: Mayor and Queloz's discovery paper for 51 Pegasi B was published in *Nature* after being revised in response to the three referees' reports. The revised paper included the results of their second eight-night observing run in September, further strengthening their already strong case for the planet. Mayor and Queloz noted that their discovery had since been confirmed by Marcy and Butler and also by a joint team from the High Altitude Observatory in Colorado and the Harvard-Smithsonian Center for Astrophysics. 51 Pegasi A was definitely wobbling like it had a planet. Three groups now agreed about that.

The question of whether the velocity variations were caused by stellar pulsations rather than by planets was answered in the negative—

there was no indication that a star like 51 Pegasi A could pulsate at the level needed to explain the observations. Furthermore, a colleague had monitored the brightness of 51 Pegasi A for several weeks in September and had found no evidence of rhythmic brightening and dimming with a 4.23-day period. While Mayor and Queloz admitted that even these observations could not quite rule out low-level pulsations in 51 Pegasi A, they referred to forthcoming observations of 51 Pegasi A by the European *Hipparchos* satellite as being much more definitive. The *Hipparchos* results could not be published without the approval of the *Hipparchos* Science Team, however, so Mayor and Queloz could only ask the reader to be patient and trusting.

Mayor and Queloz also pointed out that they had possible evidence for a *second* low-mass companion in orbit around 51 Pegasi A, with a much longer orbital period than 51 Pegasi B. Latham's quick look at his 51 Pegasi data thus seemed to be on the right track.

The *Nature* paper was accompanied by the fanfare of two commentaries written by Gordon Walker and by Adam Burrows and Jonathan Lunine. Walker pointed out that his team's recently concluded search for extrasolar Jupiters was limited to looking for orbital periods longer than 40 days; the Canadians thus may not have found 51 Pegasi B even if 51 Pegasi A had been on their short list of target stars. Walker hypothesized that there might be a whole population of gas giant planets on short period orbits, waiting to be found. He cautioned, however, that pulsations of 51 Pegasi A might yet ruin the planetary interpretation.

Burrows and Lunine presented the surprising result that a gas giant planet could be stable for billions of years in 51 Pegasi B's uncomfortably close orbit to its star. They considered the various mechanisms that would remove gas atoms from the planet and concluded that in spite of a surface temperature estimated at about 1,000 degrees Centigrade (a temperature seemingly too hot to allow gases to be retained), 51 Pegasi B would have lost only a small amount of gas over its 8- to 10-billion-year lifetime. The gravitational pull of a Jupiter-mass planet on its gas is far too strong to permit much mass loss, and it would be even stronger if 51 Pegasi B had started out its life as a more massive brown dwarf star. The possibility that 51 Pegasi B was the stripped-down remnant of a brown dwarf or low-mass star thus seemed unlikely to be the case; 51 Pegasi A was unable to consume its companion at anything close to the rapid rate at which the Black Widow pulsar was devouring its mate.

November 30, 1995: Kulkarni's Caltech group (Takashi Nakajima, Keith Matthews, and Ben Oppenheimer) and David Golimowski and

Samuel Durrance of the Johns Hopkins University published their discovery paper for the brown dwarf companion to Gliese 229 in *Nature,* followed the next day by the Caltech group's publication in *Science* of the remarkable spectrum showing the presence of methane in the cool atmosphere of Gliese 229 B. Both *Nature* and *Science* accompanied the articles with appropriate commentaries pointing out the importance of at last having found an object that most certainly fell in the netherworld between stars and planets, an undisputed cool brown dwarf star, an object whose existence had been predicted for decades but had proven to be nearly as hard to find as the first extrasolar planet. A front page story in the *Washington Post* proclaimed the discovery of a "super-planet."

The spectrum of Gliese 229 B had been shown in October at the same Florence, Italy meeting where Mayor and Queloz had revealed the existence of 51 Pegasi B, to the gasps of the appreciative audience, but the discovery of Gliese 229 B was not widely publicized until after the two Gliese 229 B papers were published, a remarkable display of journalistic respect for embargoed papers. Science journals routinely request that mass media journalists hold off on publishing the results of papers about to be published in their pages, in order to ensure the integrity of the review process for serious scientific papers and to level the playing field for the numerous journalists competing to be the first to describe an exciting new scientific development.

DECEMBER 11, 1995:   Bob Brown, Alan Dressler, and I joined Ed Weiler and Wes Huntress for a meeting at NASA Headquarters with Daniel Goldin, ostensibly to help Goldin prepare for a speech he was planning to give to the American Astronomical Society meeting in January 1996.

Goldin had been briefed the month before by Charles Elachi about the process and outcome of the ExNPS road map effort. The ExNPS road map had the virtue of a clearly stated science goal of known interest to the public (to discover and characterize extrasolar planets) that could be achieved through a series of measured steps. The scientific and technical knowledge acquired along the way would be used to guide the succeeding steps. The first steps were already underway (the ground-based searches), and the Keck adventure was about to begin. The expectation was that our galaxy was riddled with planetary systems, and the ExNPS plan would either find them or prove conclusively that they were not there.

Goldin was thoroughly pleased with Elachi's presentation, so pleased in fact that he told us that he was going to have NASA's astrophysicists create their own road map. Elachi would soon find himself leading yet another road map effort, this time dealing with NASA's plans for continuing the exploration of our Solar System.

Elachi's ExNPS road map had certainly caught Goldin's attention and earned his approval. The discoveries of the first Jupiter-mass planet and the first cool brown dwarf gave further impetus to the ExNPS road map. Surely the ExNPS road map was destined to become reality.

As usual, there was only one little problem: money. With NASA expecting to receive at best a flat budget through the year 2000, any new projects like ExNPS could only occur when some other major project was finished, thereby freeing up its annual allocation of funds. The competition would be fierce to start anything new at all.

The centerpiece of the ExNPS road map was a large space infrared interferometer, capable of detecting and characterizing extrasolar Earths and planned for a launch in 2010. This launch date was likely to produce severe trouble with the general astronomical community, however, which had its own expensive wish list.

The *Hubble Space Telescope* was nominally going to be turned off in 2005, potentially leaving astronomers without the magnificent space telescope from which they had become accustomed to receiving cutting-edge astronomical results and funds to support their research. Dressler's HST&B committee was charged with solving the problem of what to do after *HST,* and they settled on a successor called the Next Generation Space Telescope (NGST), an infrared telescope 4 meters or more in diameter that would be capable of imaging primeval galaxies in the process of birth. As a result of the expansion of the universe discovered by Edwin Hubble, our galaxy is moving away from these newborn galaxies at such high velocities that their primeval light would be redshifted to infrared wavelengths, necessitating the use of a giant infrared space telescope to capture their faint light.

If *HST*'s life was prolonged by five years or so, and if NASA could get NGST up and running by 2010, then the astronomers would notice little or no gap in their ability to make forefront observations and obtain the funding needed to do the research. But that would mean that NGST and the ExNPS interferometer would be competing against each other for a launch around 2010—and it was doubtful that NASA would be willing or able to pay for the development of two major space telescopes at the same time. Something would have to give.

DECEMBER 29, 1995: In an interview published in the *Chicago Tribune,* Marcy and Butler hinted that they were beginning to find their own evidence of a second planet in the 51 Pegasi system, just as Latham and Mayor and Queloz had. Marcy and Butler's observations were starting to point toward a second planet located close to the Earth's distance from the Sun or a bit farther out, with an orbital period of one to three years. Marcy was beginning to believe that 51 Pegasi harbored an entire planetary system.

Finding a system of several very-low-mass companions would be a nearly conclusive argument that a planetary system had been discovered. Multiple-star systems do exist, but the only ones known to be stable for long periods of time have a hierarchical structure, with increasingly larger separations occurring as one ascends the hierarchy. That is, a pair of close binary stars may be a part of a triple-star system, with the third star orbiting at a much greater distance from the binary than the distance between the close binary stars themselves. No stellar systems have been found in which several low-mass stars orbit around a much more massive star (a possibly stable configuration, but one that has never been observed). Such an orbital configuration seems to be reserved for planetary systems, which form out of protoplanetary disks after a central star has already largely formed and thus acquire their mutual orbits about the star.

It was beginning to look like 51 Pegasi would set the record not only for having the first extrasolar planet, but also the first extrasolar *planetary system* around a normal star. If Wolszczan's neutron star PSR1257+12 could have a system of planetary-mass objects, why not a solar-type star like 51 Pegasi?

JANUARY 1996:   The January issue of *Icarus* contained a paper by George Wetherill exploring the likely outcomes of the terrestrial planet formation process around other stars. Wetherill had computed 500 new models of the final phase of planet formation, when lunar-sized bodies smash together in cataclysmic collisions to form Earth-like planets.

Wetherill found that planetary accumulation proceeded in much the same way even around stars with masses significantly larger or smaller than that of the Sun. An Earth-like planet at an Earth-like distance was a typical result. However, because stars with different masses give off different amounts of starlight, the chances that an extrasolar Earth was located at the right orbital distance to evolve and sustain life were poor, unless the star was similar to the Sun. Lower-mass stars are cooler and give off less starlight than the Sun. As a consequence, their Earths tended to be frozen worlds, whereas higher-mass stars tended to have hot

Earths, where any oceans would have boiled away long ago. Because liquid water is believed to be required for the existence of life, at least life of the form with which we are familiar, Wetherill's calculations suggested that there might be something special about the habitability of terrestrial planets around solarlike stars—like Goldilocks, the Earth had found a spot that was not too hot and not too cold, but just right.

Wetherill's calculations also shed light on the identity of 51 Pegasi's companion. With a mass most likely between that of Jupiter and Saturn, it was tempting to describe 51 Pegasi B as a gas giant planet. However, Mayor and Queloz's indirect detection of 51 Pegasi B had provided no direct proof that the object was composed primarily of hydrogen and helium like Jupiter and Saturn—all they were able to provide was a lower limit on the mass. Maybe 51 Pegasi B had no significant hydrogen and helium at all, but instead was a massive hunk of silicate rock and iron metal, a "super-Earth." After all, the close proximity of 51 Pegasi B to its star had presented severe problems with understanding the formation of 51 Pegasi B at its present location and had forced theorists to invoke orbital migration to save their basic theory of planet formation and their sanity. If 51 Pegasi B was a super-Earth, maybe it had formed right there, next to its star, and that was that.

The new calculations threw an ocean of cold water on the idea of making a super-Earth at 51 Pegasi B's distance. Wetherill's simulations showed that the most massive terrestrial planets tended to form well away from their star, close to the Earth's distance from the Sun, and not 20 times closer, at 51 Pegasi B's orbit. Even if the protoplanetary disk was cool enough to allow rock and iron planetesimals to form very close to the star, which was extremely unlikely if not downright impossible, there would not be enough planetesimals around to make even an Earth-sized planet at 51 Pegasi B's orbit, much less a Jupiter-mass behemoth—there was simply not enough room in the innermost disk to put all the necessary planetesimals. This would be similar to trying to build a gigantic airport in the center of a city instead of in the distant exurbs, where land is plentiful.

If one tried nevertheless to crowd in enough rock and iron planetesimals to make a super-Earth right next to 51 Pegasi A, then the rest of the disk would be expected to be so overcrowded that it would have a mass much greater than that of 51 Pegasi A itself, a patently ridiculous situation. If an orbiting disk finds that it is much more massive than its own star, most of the disk's mass is going to end up being added to the star, not being made into ever-larger planets. There was no way that 51 Pegasi B could have formed as a super-Earth right next to its star.

Carl Sagan sent letters to Wetherill and to me the month before, calling our attention to a paper he published in 1977 with his Cornell colleague, Richard Isaacman. Using a simple computer code to model planet formation by collisions and gas accretion, Isaacman and Sagan had predicted that giant planets could form *inside* the orbits of terrestrial planets, reversing the ordering found in our Solar System. Isaacman and Sagan found this circumstance to occur when the planet-forming disk had a very unusual distribution of mass, with nearly all the disk's mass being piled up right next to the star and therefore available to make one super-planet well inside Earth's orbit, reminiscent of 51 Pegasi. Isaacman and Sagan downplayed the significance of the model in their 1977 paper, however, stating that whether "such a planetary system can form at all is highly questionable." At any rate, Isaacman and Sagan's super-planet orbited 10 times farther out from its star than 51 Pegasi B did, showing the difficulty of explaining the formation of 51 Pegasi B in place even in this extreme case.

51 Pegasi B thus still seemed to have to be a gas giant planet that had gone on a walkabout. But the big question was, How rare was 51 Pegasi B? The spectroscopic method employed by Mayor and Queloz is preferentially sensitive to finding massive planets on short period orbits—planets like 51 Pegasi B are by far the easiest to find with Doppler spectroscopy. Was 51 Pegasi B a loner, a freak of the universe, a mistake that could be safely ignored in the race to find habitable extrasolar Earths? Or was the Solar System the freak, with its Earth precisely located in the habitable zone, protected from cometary impacts by a benevolent Jupiter? Where were the extrasolar Jupiters that orbited much more slowly, out where they belonged, leaving some room for extrasolar Earths?

15

THAT'S NO PLANET,
THAT'S MY DWARF

*Scientists have odious manners, except when you prop up their
theory; then you can borrow money of them.*

—Mark Twain (1835–1910)

JANUARY 16, 1996: Daniel Goldin's airplane arrived too late to allow
him to have dinner with the dozen astronomers waiting at the Biga
Restaurant in San Antonio, Texas, but he joined us for coffee and dessert. Goldin was in town to deliver a major address the next afternoon
to the American Astronomical Society, in which he would lay out his
vision for NASA's space science goals for the next several decades. We
were hoping that Goldin would firmly commit NASA to following slavishly the ExNPS road map. The dinner had been organized in order to
give Goldin a chance to meet with the Society's top officers, with some
of the scientists who participated in the ExNPS road map effort, and
with Geoff Marcy, whose mid-October confirmation of the existence of
51 Pegasi's planet had sparked so much hope that a new era in planet
searches had really begun.

A press conference was scheduled for 8:30 the next morning, at
which Marcy was expected to announce formally his confirmation of
51 Pegasi's planet and tell us about the status of the hints he and Butler
had found for a second planet in the 51 Pegasi system. At least, that is
what Marcy and Butler's abstract promised, though it had to be written
in late October in order to meet the abstract deadline and so the abstract could already be out of date in this suddenly fast-breaking field. I
had heard that afternoon from Steven Maran, the Society's press officer,

that Marcy would depart from the prepared script and talk about something even more exciting. Marcy had kept quiet all day long at the meeting and at the dinner party that night, but he opened up after Goldin arrived. Marcy then confided in Goldin and the rest of the dinner party what he would be presenting at the next day's press conference. Marcy and Butler were about to win the next major leg of the race to find extrasolar planets.

JANUARY 17, 1996: The press conference started first thing in the morning at San Antonio's Palacio del Rio Hilton. The press conference had to be moved to a much larger room than usual because of the crowd of reporters expected to attend. Even still, the room quickly filled, with people standing in back to get a better look. Several crews were prepared to videotape and film the press conference, and their hot, bright lights trained on the speakers sitting at the table up front gave the strong impression that something unusual was going to happen.

Christopher Burrows led off, with the announcement that he and his *HST* team had found a warp in the celebrated Beta Pictoris dust disk. Rather than being perfectly flat, as would be expected for a disk of dust grains that had had many opportunities to collide with each other, a new *HST* image of Beta Pictoris had revealed that the inner portions of the disk were bent away from the disk's midplane in opposite directions on either side of Beta Pictoris. Either some rogue star recently had passed close to Beta Pictoris and stirred up its dust disk, which was highly unlikely, or else there was something big inside the dust disk that was causing the warp. A planet or brown dwarf orbiting slightly out of the plane of the dust disk would do the job. A variety of masses would be capable of causing the warp, depending on the orbital distance of the massive body. In fact, a Jupiter-mass planet orbiting at Jupiter's distance from the Sun would do just fine, thank you. Burrows had found indirect evidence of a companion to Beta Pictoris, a possible gas giant planet that could explain not only the warp found by *HST,* but also earlier evidence that the inner regions of the disk were so depleted in dust that it looked like a fastidiously tidy planet had been at work, cleaning up the place.

At a normal press conference, reporters might be expected to rush off to write and file their reports of such an exciting discovery, but this was no normal press conference. The journalists stayed in their seats, expectantly waiting, hoping for more.

Ted Kennelly of the High Altitude Observatory spoke next and told us that while his team's spectroscopic measurements had yielded the second confirmation of the reality of 51 Pegasi B, their two years of data showed no evidence for a second planet, contrary to the preliminary results of Latham, Mayor, and Marcy. According to Kennelly's observations, there could be no second planet any more massive than 1.6 Jupiter masses orbiting with a period less than about a year; any second planet would have to be less massive than that, or on a longer-period, more distant orbit. That still left a lot of interesting possibilities, though—*none* of the planets of our Solar System would have been detected by Kennelly's group.

The University of Victoria's Ana Larson reiterated the results of the Canadian group's long-term survey—12 years, 0 planets. The University of Arizona's Tristan Guillot reassured us that 51 Pegasi's planet would not be expected to evaporate away in spite of its surface being hotter than a furnace. Then I argued that there was no way to understand how a gas giant planet could have formed at 51 Pegasi B's location. By luck, the previous October I had finished analyzing a dozen models of protoplanetary disks with widely different but plausible characteristics, and none of them yielded a disk cool enough to allow water ice to exist as a solid anywhere close to 51 Pegasi B's orbit. 51 Pegasi B could only have formed much farther out and then migrated inward.

At that point the warm-up acts were over. The reporters had been getting anxious for the main act to start, like a restless crowd at a rock concert. Fittingly, Marcy was the last on the program to speak, with a presentation cryptically entitled "New Planets." Did he mean 51 Pegasi's new planets, or something else?

Marcy described the confirmation of 51 Pegasi B and mentioned that by now he and Paul Butler had amassed enough data to make a more definitive statement about the possibility of there being a second planet in orbit. Surprisingly, Marcy agreed with Kennelly that there was no evidence for a second planet around 51 Pegasi. Marcy and Butler's earlier hints of a long-term trend in the velocity of 51 Pegasi had not been borne out by more observations, even with their high-precision spectrometer. While the reality of the 4.23-day-period planet was secure, 51 Pegasi's possible planetary *system* began to dissolve.

But that was not all, not by a long shot. Marcy went on to tell the hushed audience about observations he and Butler had made of two more solar-type stars, called 47 Ursa Majoris and 70 Virginis, located 46 and 80 light years away from Earth in the Big Dipper and Virgo constellations, respectively. These two stars were part of the original sample of 120 stars that Marcy had begun monitoring back in 1988. They had already taken seven years of data on the two stars but had not bothered to analyze the data until about a month ago, after Mayor's discovery of the 4.23-day period of 51 Pegasi B suddenly opened everyone's eyes to the possibility that Jupiter-mass planets might be found on ridiculously, unthinkably, impossibly short-period orbits.

After Mayor and Queloz's announcement, Marcy and Butler began to work through the backlog of data on their 120 stars. To improve the accuracy of their Doppler velocity method, they measured the minute shift in wavelength of thousands of lines in the spectra of each of their target stars. By taking the average of several thousand Doppler shifts, Marcy and Butler could largely eliminate any random noise in the data for each star. Because of the huge number of lines being compared, the job of computing the Doppler shifts required serious computer power. Marcy and Butler managed to secure time on six Sun workstations and began crunching through their data in earnest. Butler ended up doing most of the work in his office at UC Berkeley, as Marcy continued to perform his teaching duties at nearby San Francisco State University.

By the time of the San Antonio meeting, Marcy and Butler had managed to look at 60 of their 120 stars. Butler had found two stars that were wobbling back and forth like mad, first 47 Ursae Majoris in early December, and then 70 Virginis in late December. The wobbles were so pronounced that there could be no doubt that they had bagged two more planets. While almost bursting with excitement and enthusiasm, Butler and Marcy managed to keep their two planets a secret for several more weeks, saving the news for the delicious opportunity afforded by the upcoming American Astronomical Society meeting.

Marcy told the press conference that 47 Ursae Majoris was orbited by a planet with a mass at least as large as 2.4 Jupiter masses, moving on a nearly circular orbit with a period of three years. That orbital period meant that 47 Ursae Majoris B was located at about twice the Earth-Sun distance, less than half of Jupiter's distance from the Sun. While 47 Ursae Majoris B was a little on the heavy side for a gas giant planet, compared to Jupiter and Saturn, at least its orbit was fairly far away from its central star, at a much more normal distance than 51 Pegasi B's close orbit.

Marcy and Butler had found a giant planet with an orbit considerably more similar to that of Jupiter than Mayor and Queloz's planet. 47 Ursae Majoris B's orbital radius of twice the Earth-Sun distance was two thirds of the minimum orbital radius for a giant planet suggested by my models, but now that theorists had had to accept the daring idea that giant planets might migrate inward following their formation, 47 Ursae Majoris B's orbit seemed respectable, even if it might have moved inward a tad after it formed.

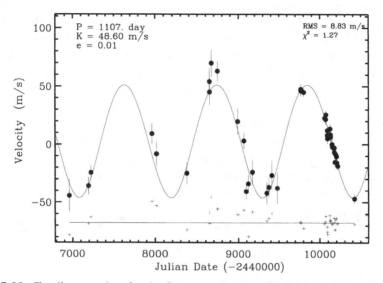

P = 1107. day
K = 48.60 m/s
e = 0.01

RMS = 8.83 m/s
χ^2 = 1.27

FIGURE 20. The discovery data for the first extrasolar planet with a more or less "normal" orbit, found around the Sun-like star 47 Ursae Majoris by Geoffrey Marcy and R. Paul Butler. The solid line fit implies a planet with a mass at least several times that of Jupiter, moving on an orbit several times as large as Earth's. (Courtesy of Geoffrey Marcy, San Francisco State University.)

Marcy then told the press conference about 70 Virginis, which their observations showed was orbited by an object with a mass no smaller than about 6.6 Jupiter masses. The orbital period was only 117 days, implying an orbital radius of about 0.4 the Earth-Sun distance, at Mercury's orbital distance. Unlike 51 Pegasi B and 47 Ursae Majoris B, the orbit of 70 Virginis B was noticeably eccentric, with an eccentricity of 0.4 on a scale that runs from 0 to 1. Marcy giddily proclaimed that 70 Virginis B was a planet, too.

Given the unknown inclinations of the orbits of 47 Ursae Majoris B and 70 Virginis B, the most probable values of their masses were 3.0 and 8.4 Jupiter masses, respectively. Marcy argued that they were both giant planets, on the basis of the expectation that the minimum mass of a brown dwarf star was between 10 and 20 Jupiter masses. Marcy and Butler thus claimed to have *tripled* the number of known extrasolar planets in a single announcement. After almost a decade of work, laboring in obscurity without any guarantee of success or reward, they had found a giant planet bonanza. And they still had 60 stars left to analyze—the vein they were following had not yet played out.

Marcy and Butler had at times been subjected to the derision of elitist astronomers, who scoffed at the importance of studying anything as mundane as a mere planetary system. Solar System comets, for example, have long been viewed as "vermin of the sky" by those astronomers whose hours-long photographic exposures of distant galaxies have been marred at times by the chance passage of a comet through their telescope's field of view. Assigning importance or priority to the incredible range of objects subject to astronomical scrutiny has always been a personal and subjective exercise, and some astronomers believe that the ultimate criterion is size or distance—the bigger the better, the more remote the more interesting. Nearby extrasolar planets would fail to pass muster on either account, but Marcy and Butler did not care. They and Mayor and Queloz had achieved something that had eluded astronomers for over 50 years: the discovery of extrasolar Jupiters, magnificent discoveries that would inevitably prompt human beings to search for and find other Earths.

Bob Brown was designated by Steve Maran to put Marcy and Butler's historic discoveries in perspective. In his eloquent way, Brown pointed out that the discoveries of the Swiss and California astronomers had completed the revolution in thought begun by Copernicus over 450 years earlier. Copernicus had dethroned the Earth from the center of the universe by showing that the Earth and the other planets revolved around the Sun. Seventy years later, Giordano Bruno asserted that the uncounted stars were suns just like our own. The inescapable conclusion of these two brilliant ideas was that other solar systems should be common in the universe, and thus that *we may not be alone.* This ultimate conclusion seemed logical and inescapable, but it had taken over three centuries for the technology to develop that could begin to *prove* the point. In 1995 and 1996, Mayor, Queloz, Marcy, and Butler had proven without a doubt that planets truly did exist around other stars like the Sun. Humankind had taken a tremendous step forward in its awareness of the universe, a step as significant as Neil Armstrong's first step onto the Moon in 1969.

The press conference ended, but the reporters could not yet disappear to work on their stories. The press conference was immediately followed by a special session with invited talks by, of all people, Geoff Marcy, and Arizona's Neville Woolf, who worked with Roger Angel at the University's Mirror Lab on designing mammoth new telescopes. The special session was devoted to the search for extrasolar planets and, incredibly, had been organized by Bill Cochran in the summer of 1995, well before anybody knew that Mayor and Queloz would revolutionize the search for new worlds in October of 1995, and well before Marcy's key role in the revolution would become clear.

After the flurry of the press conference and the special session was over, after the television lights had been turned off and the Salon del Rey room emptied for lunch, a bittersweet realization emerged. It could well have been Marcy and Butler who made the first discovery of an extrasolar planet, instead of Mayor and Queloz. Marcy and Butler had unknowingly let the evidence for their two new discoveries sit on their computer disks and tapes for years, unanalyzed, giving newcomers Mayor and Queloz the chance to win the race for the first extrasolar planet, albeit a planet on a strange orbit. With the benefit of hindsight, it was clear that Marcy and Butler easily could have beaten Mayor and Queloz to the punch if they had only analyzed their data as they made the observations.

JANUARY 17, 1996:   At 1:45 P.M., several thousand astronomers gathered in a large exhibition hall next to the Palacio del Rio Hilton to listen to the NASA Administrator's vision of the future. Through its array of space-based observatories, NASA had become a lead player in American astronomy, rivaled in output only by the NSF's ground-based observatories. When Dan Goldin spoke, everyone would listen.

Goldin reassured the audience that NASA's next two Great Observatories, following the *Hubble Space Telescope* and the *Compton Gamma-ray Observatory,* were on track and would be completed—the Advanced X-ray Astrophysics Facility and the Space Infrared Telescope Facility would be built in spite of NASA's budget problems. He stated, though, that he would resist the temptation to concentrate only on building these long-anticipated telescopes, to not start anything new. Goldin wanted NASA to think not only about what it would do in the next five years, its usual planning horizon, but also in terms of what NASA wanted to be doing 25 years from now. With such a long-term perspective, NASA could never afford to tread water contentedly—NASA must continually strike out for the next shore.

In Goldin's mind, the next shore was, to a large extent, the newly created Origins theme, with its emphasis on the formation of galaxies, stars, and planets. Goldin admitted to being personally excited and even emotional about the possibilities. He thought the ExNPS road map had a funny name but was otherwise terrific. Goldin envisioned the successive steps of the ExNPS road map program leading from the ground-based Kecks, through the first space interferometer, and to the space infrared interferometer that would find terrestrial planets.

But Goldin did not stop there. He went on to restate the dream he had advanced at his speech to the American Geophysical Union in 1994: the goal of obtaining high-resolution images of nearby Earths that showed continents and oceans, ice caps and clouds. Goldin chided those who had laughed at his dream, saying that they had simply not allowed their imaginations enough freedom. For example, _HST_ had been designed to function at the Earth's surface, where gravity dictated that the primary mirror must be a thick, heavy sheet of glass. In the microgravity environment of Earth orbit, such a rigid design was incredible overkill—there was no need for a space telescope to have a mirror that strong, massive, and expensive. Goldin challenged technologists to come up with radical ways to build truly immense telescopes in space, telescopes capable of not only finding new Earths, but of revealing details of their surfaces and atmospheres.

The San Antonio audience applauded long and loudly. Those middle-aged astronomers who hoped to witness the discovery of extrasolar Earths in their lifetimes could hardly have asked for a better NASA Administrator than Dan Goldin. Goldin had turned out to be a far bigger dreamer than _anybody_ on the ExNPS road map teams.

JANUARY 18, 1996: Marcy and Butler's discoveries were reported in stories on the front pages of the _New York Times_ and the _Washington Post;_ the front cover of _Time_ magazine would carry the story a few weeks later. Both newspaper stories focused, however, on Marcy's assertion at the San Antonio press conference that the new planets could contain liquid water and thus were capable of supporting life. Marcy was reaching for the brass ring, hoping to gain credit for finding not only two more extrasolar planets, but for the first _inhabitable_ planets—51 Pegasi's planet was too incredibly hot to be inhabited by anything resembling terrestrial life.

Marcy had stated in San Antonio that 70 Virginis's planet would have a surface temperature of about 185 degrees Fahrenheit, well below the boiling point of water. Orbiting farther away from its star, 47 Ursae Majoris's planet would have a surface with a temperature that was far below the freezing point of water, but Marcy noted that if the analogy with Jupiter was correct, there would be a layer farther down where the temperature was warm enough for water to be liquid.

Liquid water is a prerequisite for life on Earth, and anywhere there is liquid water on Earth, even if only for a few months a year, there is also life. Organic molecules (such as methane) are known to be abundant in gas giant planets, and these organics can provide the building blocks for assembling increasingly complex molecules. Given gentle heating of this primordial soup for a billion or two years, Marcy and Butler hinted, and *voila,* you have *life.* Their vision of conceivably having found abodes suitable for alien life was heady stuff indeed.

However, the chances for life existing on 47 Ursae Majoris B and 70 Virginis B were greatly restricted by the likelihood that both planets were gas giants, without a solid surface like the Earth. The surface of a gas giant planet is the top of a roiling atmosphere, a gaseous pressure cooker, where the ingredients for life would constantly circulate up and down through regions of high and low temperature and pressure. An organic molecule formed at one level might well be destroyed at a different level of the atmosphere. While it would be foolhardy to say that life could *never* evolve on a gas giant planet, it should be noted that NASA has no plans to search for primitive life on Jupiter or Saturn—the best place to look for extraterrestrial life in our Solar System is Mars, a rocky planet where liquid water once flowed across the surface and where water still exists in some abundance, frozen in the planet's polar ice caps.

47 Ursae Majoris B and 70 Virginis B would be poor prospects for life as we know it. However, if these two new planets had systems of moons, similar to those around Jupiter and Saturn, that would be a different matter—Jupiter's icy moon Europa and Saturn's giant, methane-rich moon Titan are dark horse candidates for having experienced at least some organic chemical evolution. Marcy and Butler had suggested this possibility as well but were unable to present any evidence for the existence of extrasolar moons other than the analogy with our Solar System. Marcy and Butler had reached prematurely for the Holy Grail, the ultimate prize in the race to find other planetary systems, but few scientists were ready to award them that achievement.

JANUARY 19, 1996: I returned to my office at DTM to find the telephone's message light blinking, unread e-mail messages filling the screen of my computer, and a stack of pink slips about telephone calls that needed to be answered. The media had gone wild about Marcy and Butler's story—the Cable News Network wanted me to do a live interview, and even Oliver North wanted a live interview for his Washington-area radio program. I could only imagine what Geoff Marcy and Paul Butler were experiencing, the ones who really deserved the attention, if someone as peripheral to the story as me was getting inundated.

Unfortunately, I was extremely ill with a gastrointestinal infection that had made my flights home from San Antonio the day before an absolute nightmare. Having made it into my office, I could not muster the strength to return the phone calls and do the interviews. I tried to gather my energy by resting on the couch in my office, but then the Oliver North show insistently called again. I asked George Wetherill to do the North interview for me, gave him the stack of pink slips, and then I headed home and got back in bed.

One of the e-mail messages was from Leslie Sage, who wanted me to write a commentary about the new planets for *Nature* as soon as was humanly possible. I trudged back into DTM on Sunday, January 21 and tackled the job of making sense of what Marcy and Butler had found.

The popular press had rather uncritically accepted Marcy and Butler's claim that both of the new objects were "planets," when it was apparent from the conventional wisdom that they were only half right. While 47 Ursae Majoris B was indeed likely to be a gas giant planet like Jupiter, 70 Virginis B was most likely a brown dwarf star. The key discriminator was not mass, but the shape of their orbits—unmolested planets should have nearly circular orbits, while brown dwarf stars with orbital periods greater than a couple of weeks should have eccentric (elliptical) orbits, at least if the conventional wisdom about star and planet formation was right. By the orbital eccentricity criterion, 47 Ursae Majoris B was a giant planet, while 70 Virginis B was a brown dwarf star.

Before the San Antonio press conference, there was a well-defined gap between the mass of the most massive planet then known, Jupiter, and the mass of the least massive suspected brown dwarf star, HD 114762 B, which could have a mass as small as about 11 Jupiter masses. The mass of Mayor and Queloz's 51 Pegasi B, about half that of Jupiter, fit in nicely between the masses of Jupiter and Saturn. Before San Antonio, there was thus a comfortable gap between the most massive planet and the least massive brown dwarf star, a gap of at least a factor of 11. That gap implied that we might well be able to tell the difference between the two types of beasts solely on the basis of their masses.

Marcy and Butler's twin discoveries nearly eliminated that gap. Now the most massive planet known was 47 Ursae Majoris B, weighing in at about 3 Jupiter masses. Marcy and Butler had obliterated the title of the previous record holder, Jupiter. Jupiter's mass had been known ever since the 1600s, after the Italian Galileo Galilei discovered and watched four of Jupiter's moons; with a little help from the study of gravity by Isaac Newton, the moons' orbital characteristics could be used to calculate Jupiter's mass. Over three centuries later, Marcy and Butler's 47 Ursae Majoris B displaced Jupiter as the most massive planet.

By interpreting 70 Virginis's companion as a brown dwarf star rather than as a planet, the companion's mass being as small as about 6.6 Jupiters meant that Marcy and Butler had probably found the least massive brown dwarf star to date—they had set *two* records in the *same* press conference, quite an achievement even by the frenetic standards of contemporary astronomy. However, if this interpretation was correct, Marcy and Butler's two discoveries also meant that the mass gap between planets and brown dwarf stars had shrunk from a factor of 11 to a factor of about 2. This was an uncomfortably small difference, because astronomers routinely consider numbers that are within a factor of 2 of each other to be essentially the same number, to astronomical accuracy.

Orbital eccentricity thus became the key discriminator between giant planets and brown dwarfs, with the slim remaining mass gap as a supporting argument. For all we knew, though, the next discoveries that were sure to come might well produce even more massive planets and less massive brown dwarfs, eliminating the mass gap and throwing the entire identification scheme into chaos. At that point, theorists like me might have to throw up their hands and concede that the universe was more complicated than they had imagined, a chastening act that many observers would love to watch. For the moment at least, however, a dividing mass around 4 Jupiter masses fit the observational data and fit in well with my late 1980s estimate of the minimum mass of a brown dwarf star.

I finished the draft of the *Nature* commentary that afternoon and faxed a more polished version to Sage on Monday, the next day. By Tuesday *Nature* had faxed me back the galley proofs of the commentary to check for printer's errors. The piece was published about a week later in *Nature*'s February 1 issue, a compression of the usual year-long delay for scientific journal publication into less than two weeks. Extrasolar planets were hot, and the word had to get out fast. Marcy and Butler's discovery papers about 47 Ursae Majoris B and 70 Virginis B would soon be submitted but would not appear in the *Astrophysical Journal Letters* until the June 20 issue.

After faxing the manuscript to *Nature* on Monday, I rushed home to grab a suit and tie and then hurried back to DTM in order to meet a limousine that drove me to a television studio hidden on K Street Northwest in downtown Washington, D.C. Cable NBC had asked me late that afternoon to do a live television interview on its *America's Talking* nationwide show. I soon found myself sitting on an uncomfortable stool in front of a live television camera as the show's host and hostess asked me not about the new extrasolar planets, as I had expected, but about the implications for *life* on *Jupiter* of NASA's *Galileo* mission, whose probe had just sampled the upper atmosphere of the giant planet and found some peculiar results regarding water. Marcy and Butler's suggestion that their extrasolar Jupiters might harbor liquid water and hence some sort of creatures seemed to have helped spark the suspicion that Jupiter itself might be inhabited. I tried to steer the interview toward the new planets instead, but the host and hostess wanted only to talk about the possibility of life on Jupiter, a subject for which my qualifications as an "expert" were minimal to nonexistent, at best. I suddenly realized I was there under false pretenses, but it was much too late to do anything about it—the red light was shining on the studio camera, meaning *you are on*.

I kept a straight face and carried on with the interview as best I could, and then escaped to the darkness and safety of the limousine for the ride back to DTM. In the end, the limousine's chauffeur performed the interview I had been expecting, asking me many knowledgeable and thorough questions about the new extrasolar planets and the implications for intelligent life beyond our Solar System.

16

WHERE'S THE BEEF?

The Earth is the cradle of mankind. But one does not live in the cradle forever.

—Konstantin Eduardovich Tsiolkovskii (1857–1935)

In the short space of four months, the Swiss and the Californians had proven that at least two extrasolar giant planets existed. The centuries-long wait for the first extrasolar planets around normal stars was over. But where were the stars that could be orbited by an entire retinue of planets, that might even hide an *Earth-like* planet?

51 Pegasi had a Jupiter-mass planet, but the planet's short-period orbit meant that it must have created havoc among any Earth-like planets it passed by during its inexorable inward migration. Not only that, but the outer atmosphere of 51 Pegasi was known to be enriched in metals like iron, compared to the Sun's iron abundance. Doug Lin made the gruesome suggestion that perhaps that was all that remained of any Earth-like planets that 51 Pegasi might once have had—just a belch of iron lingering in the star's atmospheric breath.

We had learned that Gliese 229 A was circled by a cool brown dwarf star. The Pluto-like separation of the brown dwarf from Gliese 229 A, however, meant that it might be hard to form Earth-like planets around Gliese 229 A. Depending on its mass and orbital eccentricity, the close-by brown dwarf star Gliese 229 B might well interfere with the formation of an Earth around Gliese 229 A while providing insufficient starlight to make habitable any Earth of its own. Similarly, 70 Virginis's eccentric brown dwarf companion would surely prevent the formation of any rocky, Earth-sized planets in the habitable zone of 70 Virginis A—70 Virginis B's orbit placed it near the inner edge of the primary star's habitable zone, where the brown dwarf could be expected to do the maximum damage possible to any planetesimals trying to form a planet a little farther out.

47 Ursae Majoris looked somewhat more promising, but with its mass on the order of three Jupiters and an orbital radius just twice that of Earth's, there was no way that an Earth-like planet could form in the habitable zone of 47 Ursae Majoris—Wetherill's calculations implied that if a 3 Jupiter-mass planet was located at that distance, its gravitational tugs would disrupt the terrestrial planet formation process, restricting any surviving rocky planets to the hellish existence of fiery orbits close to 47 Ursae Majoris A.

In short, none of the systems found so far looked promising as homes for extrasolar Earths. But this situation was sure to change. Now that astronomers had shown that a number of very-low-mass companions existed around nearby stars, and by doing so had revealed exactly how to find more, there would certainly be a rush by many groups to find additional extrasolar Jupiters and brown dwarfs—short of finding the *first* of any new type of celestial object, there is nothing that an astronomer loves more than finding the second one, or one of the next dozen, or even one of the next hundred.

Because Mayor's and Marcy's spectroscopic searches worked best at finding massive planets on short-period orbits, it might take a while for Doppler spectroscopy to find Jupiter-mass planets with Solar-System-like orbits—about 12 years, in fact, the duration of the Canadian spectroscopic search. Marcy was already eight years into his search but had only been taking high-precision data for a year or so. Mayor had only two years of data with his new spectrometer, so Marcy had a tremendous head start. Cochran's and McMillan's spectroscopic searches were also in the race, along with Gatewood's astrometric search. Like Marcy, they had been searching for planets since the late 1980s, and it was about time for *somebody* to find a Jupiter on a long-period orbit, an orbit big enough to leave some room for an Earth.

MARCH 11–15, 1996: The Infrared Space Interferometry Workshop was held inside the medieval city of Toledo, Spain and focused on the search for Earth-like planets. The international conference provided a chance for the Americans in attendance to find out how their European colleagues were planning to go about finding more planets.

The European Southern Observatory's (ESO) Francesco Paresce told us about ESO's plans for searching for planets with their Very Large

Telescope (VLT), then being built in Chile: *four* telescopes, each greater than 8 meters in diameter, with more light-gathering power than even the twin Keck telescopes. Contrary to the several-year-old PSSWG rumor, ESO was still planning to combine all four telescopes into a gigantic interferometer, the Very Large Telescope Interferometer (VLTI), and might even add a few smaller outrigger telescopes to help improve the ability of the interferometer to resolve fine details in different directions. The VLTI was beginning to sound a lot like NASA's plans for the Keck interferometer. By working at infrared wavelengths, Paresce hoped that the VLTI would achieve the 10-microarcsecond accuracy that was needed to detect the astrometric wobble caused by Uranus-mass extrasolar planets. If that worked, the VLTI would be serious competition for the Kecks and might even obviate the planet-search component of NASA's first space interferometry mission—AIM/OSI. It was evident that a high-stakes hardware race was underway.

The Europeans also had their competing vision of the space infrared interferometer that would search for and characterize Earths, a project they called Darwin. However, the Europeans were no more assured of being able to fly Darwin anytime soon than NASA was of flying its own space infrared interferometer—there were many other ideas competing for a limited number of flight opportunities. Sergio Volonte of the European Space Agency (ESA) told us, in fact, that ESA's budget had just been reduced and that ESA was rethinking all of its plans. ESA was beginning to feel NASA's budgetary pain; as a result, a collaborative effort in space might be attractive, as had occurred before with *IRAS* and *HST*.

Michel Mayor told us that he still believed that there might be a second planet in the 51 Pegasi system, with a mass similar to that of 51 Pegasi B, about half a Jupiter mass. A planet of that mass on a long-period orbit would not have been detected by Kennelly and his group. Furthermore, the measurements that hinted at the second planet had been taken by Mayor and Queloz *before* Marcy and Butler began their high-precision monitoring of 51 Pegasi. During the limited period that Marcy and Butler were observing 51 Pegasi, Mayor believed, the periodic Doppler shift caused by the second planet was passing through zero, consistent with Marcy and Butler's null result. Unfortunately, 51 Pegasi was once again invisible, hiding close to the Sun, and would not be visible at night until the next summer, so the mystery of 51 Pegasi's second planet would remain unsolved at least that much longer.

While in Toledo, I received a fax inviting me to appear on John McLaughlin's *One on One* nationwide public television show. McLaughlin wanted to broadcast an Easter weekend show about life on other planets; the brand new era in extrasolar planet discoveries evidently was beginning to pop up on the radar screens of political Washington. I called McLaughlin's producer in Washington, D.C., but found out that the show's taping date conflicted with the annual Lunar and Planetary Science Conference in Houston, which to my scientist's mind obviously took priority. I naively asked the producer to reschedule the taping, not knowing that McLaughlin had once decided to drop Vice-President Albert Gore from the show for requesting a two-hour delay, forced by an unexpected meeting of Gore with Newt Gingrich and Bob Dole. After hanging up, it dawned on me that it would not exactly be the end of the world if I missed a few talks at the Houston meeting. I quickly called back McLaughlin's producer, tail between my legs, and agreed to do the show as scheduled.

MARCH 20, 1996: Ed Weiler and I arrived at the NBC studio in northwest Washington to videotape McLaughlin's *One on One* show, or, in this case, "one on two." McLaughlin was taping three shows that day, and we were the first.

McLaughlin, a Jesuit priest with a Doctor of Divinity degree, is the holy terror of Washington's political talk shows, a domineering host with the style of a Grand Inquisitor. World leaders and political pundits alike approach his shows with some trepidation. What would happen to a couple of innocent scientists? Were we going to get roasted alive like Giordano Bruno?

I had spent several hours the day before at McLaughlin's downtown offices, helping him and his staff prepare for the show, and I realized then that Weiler and I were not in mortal danger. While McLaughlin's demeanor in private is much like his public persona, it became clear that McLaughlin had a genuine interest in learning as much as he could about the new discoveries and the implications for life on other worlds. McLaughlin would be treading on our turf, slowly picking his way across the relatively unfamiliar terrain of other worlds. Weiler and I would be his guides, not his victims.

McLaughlin was fascinated by the numerical implications of the Swiss and Californian discoveries. Based on the number of stars searched to date, extrasolar planets were beginning to appear with a frequency on the order of a few percent, and the search was just starting—the true frequency of planets and planetary systems had to be much higher. While we had yet to find anything smaller than a Jupiter, if the frequency of Earths was as high as a few percent, then there might be many, many Earths in the Milky Way galaxy alone, nearly a billion. McLaughlin marveled at the realization that there are at least as many galaxies in the universe as there are stars in our galaxy, so that the universe might well contain more than a billion billion Earths. How could we possibly be alone in the universe, with that many locations where life might evolve?

Long after the videotaping was over, McLaughlin sat on the show's set and continued to ask us questions—he was still enthralled. For a program in which NASA had yet to spend any serious money, the search for extrasolar planets was roaring along in terms of public awareness, seemingly with a life and a mind of its own. Weiler could only hope that the momentum would be maintained, that NASA would indeed find the funds to implement the grand plan envisioned in the ExNPS road map.

Late in March, a final shootout occurred between the two contenders for NASA's first space interferometry mission, Mike Shao's OSI and Bob Reasenberg's POINTS. OSI had been chosen over POINTS by NASA Headquarters the year before, but Reasenberg had lodged a formal complaint with NASA over the manner in which the 1995 decision had been made, without the benefit of peer review. After 20 years devoted to his work on POINTS, Reasenberg was not about to give up easily.

NASA obliged Reasenberg's request and convened an ad hoc review panel to make the final decision, chaired by David Black at Reasenberg's request. POINTS had its second chance. However, the review panel reaffirmed NASA's 1995 edict—Shao's OSI design *was* superior to Reasenberg's POINTS for what NASA had in mind, a mission that would fulfill the goals of the Bahcall report's AIM mission and would serve as a technological forerunner for the Earth-seeking space infrared telescope, now called the Planet Finder. POINTS was revived in order to die a second time.

With OSI safely reannointed as the best design for AIM, NASA was at last ready to move forward with the first space-based element of the ExNPS road map. Goldin and Huntress put 12 million dollars into the next year's budget to get the new project moving at JPL. A new project needed a new acronym, of course, so OSI/AIM became the Space Interferometry Mission, or SIM, reflecting its dual role as a science mission and as a technological precursor to the Planet Finder. If all went well, SIM would be ready to fly in 2003.

Goldin and Huntress were willing to spend about 10 percent of NASA's annual science budget on the search for Earths, as much as 200 million dollars a year. SIM alone was expected to cost about 400 million dollars, including the cost of the launch into space. The Planet Finder could not cost much more than that, a total of perhaps half a billion dollars, if Dan Goldin had his way. Planet Finder must be capable of far outperforming *HST* in the search for planets, yet could cost only a fraction of the roughly 2 billion dollars needed to build *HST*. Similarly, NGST would also have to cost no more than a fraction of *HST*.

Dan Goldin was not just being provocative when he issued these challenges to the telescope designers—with his background in building top-secret surveillance satellites for the Department of Defense, he knew that there were several revolutionary ideas brewing for building large space telescopes. Most of the savings for building SIM and the Planet Finder would come from their basic interferometric designs, but for NGST, in particular, a cheap way to build large, monolithic mirrors would have to be found.

HST's primary mirror was rigid enough that it could be tested fully on Earth prior to being launched into space. It was thus thought to be possible to guarantee the performance of the primary mirror—assuming the ground testing was performed correctly, a stunningly incorrect assumption as it turned out. Because of the human temptation to price objects by the pound, the cost of a spacecraft is directly dependent on the weight of its instruments, so *HST*'s heavy mirror led to a hefty price tag.

Dan Goldin therefore suggested the use of very lightweight mirrors that could *only* function in space. Total mission costs could be greatly reduced, though there would be substantially increased risk that a telescope that could not be fully tested on Earth might not perform well, or at all, in space. One of the ideas for radically new, lightweight designs is a "flower petal" that unfolds after reaching the microgravity environment of Earth orbit, with each petal becoming a segment of the mirror surface. Another involves a stack of hexagonal mirrors that spreads apart in space like a hand-held fan, forming a hexagonally tiled surface like that of the Keck telescopes. A plastic bag might be inflated like a balloon once the telescope reaches Earth orbit, where it will harden into a sunshield after being exposed to the vacuum of outer space. NGST and the Planet Finder telescope might turn out to be as remarkable for their innovative designs as for the baby galaxies and extrasolar Earths they would discover and scrutinize.

APRIL 12, 1996: I received a telephone call at 3:30 P.M. from Kathy Sawyer, the *Washington Post* journalist who had been reporting on the extrasolar planet stories. Paul Butler would be giving a talk at 4:00 that afternoon at the University of Maryland, and was I going to attend? No, I was not, but I changed my mind as soon as Sawyer hinted about what Butler would be revealing. Sawyer was not able to attend herself, and she wanted to hear my reaction to Butler's talk. I jumped into my car and sped to the College Park campus, ran to the Physics Lecture Hall, and slowly walked down the steps into the Hall just as Butler was being introduced.

Butler had earned his Ph.D. degree at the University of Maryland, so he was returning in triumph, wearing the twin laurels of 47 Ursa Majoris B and 70 Virginis B. Butler was introduced as an astronomer who had just ascended to the select ranks of Herschel, Galle, and Tombaugh, those fortunate few who had found new worlds. The Maryland audience gave him a deservedly warm welcome.

Butler's talk was surprisingly short, a mere half hour. Astronomers normally carry on for a full hour or so before pausing for questions from the audience. Some then take the opportunity afforded by the audience's questions to return once again, with gusto, to their main subject, effectively stretching the talk to an hour and a half or more. Butler was much more merciful—he had something important to present, and he wasted little time in getting to the point.

Marcy and Butler had been busy since the San Antonio meeting computing the Doppler shifts for 40 more stars that they had been unable to analyze prior to the January meeting. They had struck gold again and still had another 20 stars left to analyze.

Butler announced that he and Marcy had found the signature of a planet with a mass at least 80 percent that of Jupiter orbiting around the star 55 Rho[1] Cancri. 55 Rho[1] Cancri is a normal star with a mass somewhat less than that of the Sun, located about 45 light years from Earth in the constellation of Cancer. The most probable value of the mass of the planet was exactly a Jupiter mass—Butler and Marcy had bagged *another* giant planet. The discovery put them ahead of Mayor and Queloz: The score was now 2 to 1, in Marcy and Butler's favor.

55 Rho[1] Cancri B turned out to be more similar to Mayor and Queloz's 51 Pegasi B than to 47 Ursa Majoris B: 55 Rho[1] Cancri B had a very-short-period orbit, 14.7 days long, compared to 51 Pegasi B's 4.23-day period and 47 Ursa Majoris B's three-year period. 55 Rho[1] Cancri B's period was so short that its orbital separation from its star was about one tenth that of Earth's, meaning that its orbit was less than three times smaller than Mercury's. 55 Rho[1] Cancri B would be a hot Jupiter like 51 Pegasi B, though not quite as hot, because its orbit was

twice as large and its star was somewhat cooler. Evidently, 51 Pegasi B was not the only oddball planet in the galaxy.

One remarkable characteristic of 55 Rho[1] Cancri B was its orbital period, a little over two weeks. While short compared to Jupiter's 12 years, 14.7 days was just long enough that if 55 Rho[1] Cancri A was about the same age as the Sun, then tidal forces between the planet and the star would not yet have had enough time to change the planet's orbit and force it to become circular, if it had ever had an elliptical orbit to begin with. That is, if 55 Rho[1] Cancri B had formed with an eccentric orbit, it would be expected to still have that eccentric orbit—it would still have its primordial eccentricity. 55 Rho[1] Cancri B thus provided an important test of the theoretical claim that planets should form on circular orbits, and brown dwarf stars on eccentric orbits.

Given its mass, 55 Rho[1] Cancri B looked to be a gas giant planet, but what about its orbital eccentricity? Butler did not mention the orbital eccentricity of 55 Rho[1] Cancri B during his brief talk. I waited for his talk to end, wondering the whole time if the giant planet/brown dwarf classification scheme I had advanced in *Nature* was about to come crashing down, and then popped the question. Butler replied that 55 Rho[1] Cancri B was on a circular orbit, as best they could tell. Bingo. The classification scheme had passed its first test. 55 Rho[1] Cancri B was a planet.

Butler hinted that he and Marcy had evidence that there might be a second planet in the 55 Rho[1] Cancri system but could not yet pin down its properties. That would take more time.

There were now three extrasolar giant planets. Only one of the three, 47 Ursa Majoris B, looked remotely Solar System-like; the other two must have migrated inward to their present scorching orbits. Butler and Marcy still had more data to analyze and would most likely be finding more planets, but the spectroscopic method would preferentially find Jupiters with short-period orbits. Where were the extrasolar planets with Jupiter-like orbits? Who would find the first one?

APRIL 13, 1996: Geoff Marcy sent me an e-mail about 55 Rho[1] Cancri B, raising the question of whether or not Doug Lin's mechanisms for stopping the inward migration of extrasolar planets would work for 55 Rho[1] Cancri B.

One of Lin's mechanisms relied on tidal forces between the star and planet. Tidal forces depend extraordinarily strongly on the distance between the two bodies. Increasing the distance between a star and its planet by a factor of 2 decreases the tidal forces by a factor of 64. Because 55 Rho[1] Cancri B's orbit was twice as large as 51 Pegasi B's, the tidal forces that Lin had invoked to stop 51 Pegasi B's inward migration would be extremely weak at 55 Rho[1] Cancri B's orbit.

I agreed with Geoff that 55 Rho[1] Cancri B might indeed make problems for the tidal scenario, but noted that the second mechanism proposed by Lin and his colleagues in their still unpublished *Nature* paper might work (publication of the paper was being held up by a curiously unresponsive referee). Lin had pointed out that the inward migration would also stop when the planet reached the inner edge of the protoplanetary disk, where the planet would go into a parking orbit. Steven Strom and his University of Massachusetts colleagues had found evidence for sizable inner holes in suspected protoplanetary disks around young stars, and these gaps could easily extend to 55 Rho[1] Cancri B's orbit and beyond. Perhaps 55 Rho[1] Cancri B had been moving inward at the same time that the inner edge of the disk was being eroded away, leaving 55 Rho[1] Cancri B to orbit wherever it reached the edge.

55 Rho[1] Cancri, like 51 Pegasi, did not appear hospitable to Earth-like planets. Where were the stars with Jupiter-mass planets on wide orbits, with plenty of room for Earths?

APRIL 29, 1996: I sent George Gatewood an e-mail message inquiring about a rumor I had heard a few weeks before at a meeting at the Naval Observatory in Washington. The rumor was that Gatewood had a planet of his own. I was in the process of writing an article for *Physics Today* about the new planets and did not want the article to be out of date before it was published, a real danger considering the suddenly rapid progress in this once moribund field. Gatewood obliged by sending me his abstract for the upcoming American Astronomical Society meeting in Madison, Wisconsin.

Gatewood's results dealt with the fourth closest star to the Sun, Lalande 21185, conveniently lying a mere 8.2 light years away. In 1801, the French astronomer Joseph-Jerome de Lalande had published a catalog of the location of 50,000 stars; Lalande 21185 was just one of the many stars on that list, and a faint star at that, three times less massive than the Sun. Lalande himself had narrowly averted everlasting fame in 1795 by failing to notice that the position of one of his "stars" had changed markedly from one night to the next, an occurrence that should have sounded alarm bells, especially coming 14 years after Herschel's discovery of Uranus. Only after Galle found Neptune in 1847 would it be realized that Lalande had the clues necessary to stumble on Neptune, but had ignored them.

Lalande 21185's closeness made it a natural candidate for planet searches by the astrometric method, in which closeness really counts. Lalande 21185 had been on the observing program at van de Kamp's Sproul Observatory ever since 1912. Van de Kamp himself had found in 1944 that Lalande 21185 seemed to have a wobble and so might have a companion. In 1960, Sproul's Sarah Lee Lippincott used 315 nights of data amassed over 47 years to report evidence of a companion to Lalande 21185 with a mass on the order of 10 Jupiters, moving with a period of 8.0 years.

The Sproul Observatory claim for Lalande 21185 was not to last. In 1974, the year after he and Eichhorn had deflated van de Kamp's claims for planets around Barnard's star, Gatewood published his own study of Lalande 21185, performed at the Allegheny Observatory. His two-page paper in the *Astronomical Journal* stated that his 143 exposures of Lalande 21185, taken between 1934 and 1972, revealed no sign of a wobble caused by a 10 Jupiter-mass companion with an eight-year period.

Over 20 years later, though, Gatewood had found evidence for a planet around Lalande 21185 very different from that proposed by Lippincott. Gatewood had been monitoring Lalande 21185 for eight years with his new electronic device and could now measure stellar wobbles 10 times smaller than before. His plan was to present the results as a poster paper at the Madison meeting, a relatively low-profile mode compared to the alternative of an oral presentation before an audience of astronomers. In addition, Gatewood was not planning on letting Stephen Maran, the Society's press officer, in on the secret. In comparison to Marcy and Butler, who had learned early on about the importance of informing the press about their results, Gatewood was planning to make his announcement in so subtle a way that it might be missed entirely—Gatewood had a long-period personality, you might say, compared to Marcy and Butler. After Maran's successful orchestration of the press conference at the San Antonio meeting, I thought that Gatewood needed to warn Maran, so that journalists could be alerted to what Gatewood was about to announce. I gave Maran's telephone number to Gatewood and hoped that he would give Maran a call.

He did. Gatewood had found the beef.

JUNE 11, 1996: In a discussion with reporters at the American Astronomical Society meeting, Gatewood announced that he had found as-

FIGURE 21. George Gatewood, professor at the University of Pittsburgh, Director of the Allegheny Observatory, and discoverer of the first probable planetary system, around the solar-type star Lalande 21185. (Courtesy of the photographer, Cornelia Karaffa, Pittsburgh, Pennsylvania.)

trometric evidence for a system of Jupiter-mass planets orbiting around Lalande 21185. Lalande 21185 was wobbling around like it was orbited by *several* Jupiter-mass planets, the most certain one having an orbital period of about 32 years and another one having a period of 6 years. Those orbital periods implied distances from Lalande 21185 of about 7.2 and 2.4 times the Earth-Sun distance, bracketing Jupiter's distance from the Sun. There might even be a third planet in the system, orbiting farther out, but many more years of observations would be needed to establish its existence and to nail down its characteristics.

The two Jupiter-mass planets seemed to be moving on circular orbits that were more or less in the same plane, duplicating the configuration of the gas giant planets in our Solar System. Not only had Gatewood presented evidence for what appeared to be the first extrasolar planetary *system,* Lalande 21185's planetary system looked reassuringly similar to ours: two gas giant planets on coplanar, circular orbits, well away from the central star.

The inner giant planet, like 47 Ursae Majoris B, might well prove hazardous to the growth of any terrestrial planets around Lalande 21185 A, but the outer giant planet looked like a benign giant that could protect any Earth-like planets from cometary impacts. Gatewood's evidence for the outer planet was good, but the inner planet was admittedly more dubious—in fact, if it did not exist at all, Lalande 21185 might be a considerably more hospitable locale for extrasolar Earths. There was plenty of room to build and retain terrestrial planets between Lalande 21185 A and the 32-year-period giant planet.

After the shocks of 51 Pegasi's and 55 Rho[1] Cancri's hot planets and the improved prospect provided by 47 Ursa Majoris's planet, Lalande 21185 became the best hope for a nearby Solar System analog. Its closeness made Lalande 21185 the natural choice for subsequent planet searches as well as attempts to characterize Gatewood's planets.

A flurry of newspaper stories reported Gatewood's discovery. Gordon Walker, writing in *Nature* a few weeks later, noted that Lalande 21185's innermost giant planet was on the verge of being able to be imaged in the near infrared by a variety of ground-based telescopes and by *HST*, after an infrared camera was installed in early 1997. If these planets could be imaged, their brightness could be measured, and from that their size and density could also be estimated, giving final proof that a large, low-density gas giant planet had been found.

17

THE PLANET-A-WEEK CLUB

We have now entered a new era of human history, in which we are able to detect the planets of other suns. Worlds of Jupiter mass have been found around Sunlike stars, but not yet any like the Earth.

—Carl Sagan (1934–1996), June 9, 1996

Paul Butler had joked that with Doppler spectroscopy having been proven to be an efficient means for discovering extrasolar planets, the world was about to get a subscription to the Planet-a-Month Club. He was almost right. It turned out to be more like a subscription to the Planet-a-Week Club.

JUNE 16–18, 1996:   A conference on planet formation around binary stars was held at the State University of New York's Stony Brook campus on Long Island. The conference organizers, Stony Brook's Michael Simon and UCLA's Andrea Ghez, had conceived of the conference well before the discoveries of 1995, motivated in large part by their observations of protoplanetary disks around single and binary young stars. In the interim, several genuine extrasolar planets had been found, but seemingly not about binary stars.

Geoff Marcy was the lead-off speaker, and he had much to say. First, he pointed out that he still had no evidence for a second companion to 51 Pegasi, contrary to Mayor's belief. Then he said that 55 Rho[1] Cancri appeared to be a member of a binary with a period of about 30,000 years, a period so long that it was hard to be sure if the two stars really revolved about each other. They were both moving in the same direction in the sky, however, and so were probably bound together. Thus, 55 Rho¹ Cancri B was probably the first planet in a *binary star system*—another first for Butler and Marcy and an appropriate announcement, considering the conference's theme.

Marcy had excluded binaries with separations less than about 30 to 50 times the Earth-Sun distance from his sample of stars because of the expected hard time a planet would have forming with two stars simultaneously telling it where to go; but 55 Rho¹ Cancri was so far apart from its probable stellar companion that planets could still form around it, oblivious to the presence of the other star. Marcy and Butler had thus proven that planets could form around the stars in wide binary pairs. In fact, they were also monitoring the 61 Cygni binary, for which Strand claimed a 16 Jupiter-mass "planet" in 1943.

In addition, Marcy provided more information about Butler's hunch that 55 Rho¹ Cancri had not one planet, but two: Their data now suggested a second planet with a mass greater than 5 Jupiters, moving on an orbit even farther away from 55 Rho¹ Cancri than Jupiter's from our Sun. With such a large mass, 55 Rho¹ Cancri C sounded like it might be more of a brown dwarf than a planet, and it would be hard to estimate its orbital eccentricity with the available data. If 55 Rho¹ Cancri C turned out to be real, and a planet to boot, then Marcy had announced the *second* planetary *system,* just a few days after Gatewood's announcement of the system around Lalande 21185.

Marcy still was not through. He and Butler had found another suspected hot Jupiter, an object with a minimum mass 3.7 times that of Jupiter, orbiting every 3.3 days around the normal star Tau Bootis. Tau Bootis, 60 light years away and about 20 percent more massive than the Sun, is a member of a visual binary system with a separation about 240 times greater than that of the Earth and Sun. The planet's circular orbit placed it about eight times closer to its star than Mercury's orbital distance, so close that tidal forces were certain to have circularized its orbit: the primordial eccentricity of Tau Bootis B was lost long ago, making its identification problematical. If it was a planet, then it was certainly a hot Jupiter, like 51 Pegasi B and 55 Rho¹ Cancri B. Its mass, though, put it squarely in the gray area between giant planets and brown dwarfs.

Tau Bootis had been suspected of having a companion in the 1991 survey by Duquennoy and Mayor. Mayor had fired off an e-mail message on June 14, before Marcy's talk, alerting people to that fact. This was in response to an e-mail about Marcy and Butler's Tau Bootis results sent to the community on June 13 by France's Jean Schneider, who was maintaining an Internet site with extrasolar planet information.

Fortunately, Mayor's 15 years of data agreed well with Marcy's—they both derived a 3.3-day period for the companion.

The detection of Tau Bootis B looked solid. A *second* binary star system had been proven to have a planet, and binary stars are as common in the galaxy as grains of sand on a beach. The Stony Brook conference was an unequivocal success.

JUNE 23, 1996: One week later, Geoff Marcy was at it again. The Astronomical Society of the Pacific was meeting at the Westin Hotel, in Santa Clara, California, and Marcy announced there that he and Butler had found yet another hot Jupiter. This one orbited around the normal star Upsilon Andromedae, with a mass at least 60 percent that of Jupiter, and its 4.6-day period implied a blisteringly hot orbit about 19 times smaller than the Earth's. Upsilon Andromedae B's orbit was roughly circular, as had to be the case for such a short-period planet. The primordial eccentricity was thus again unknowable, but based on the mass alone, Upsilon Andromedae had a gas giant planet.

With this latest discovery, it was becoming clear that widely separated binary stars were amenable to having gas giant planets: Upsilon Andromedae had a binary star companion even more distant than those of 55 Rho[1] Cancri and Tau Bootis.

Marcy was beginning to run up the score in his favor, but it was Mayor's turn at bat next.

JULY 1–5, 1996: The Fifth International Conference on Bioastronomy was being held on the southern Italian island of Capri. Michel Mayor presented his data for a number of new objects with masses roughly in the range of 18 to 40 Jupiter masses. Because Mayor was still in the process of refining his observations, he showed the list of new objects for just a few moments, enough time for the audience to see that the objects really existed, but not long enough for the names to be copied. In the competitive world of extrasolar planet discoveries, Mayor did not want to make things too easy for the competition—arch rival Paul Butler was also at the conference. A list of 10 brown dwarf companions to nearby stars would be revealed in a few months, after Mayor had completed his analysis.

Mayor's new objects were on eccentric orbits, clinching their mass-based identification as brown dwarf stars. With Mayor's new objects, a clear pattern was beginning to emerge—the "planets," with masses less than about 4 Jupiter masses, all had circular orbits, while the "brown dwarf stars," with masses between about 4 and 80 Jupiter masses, were all on eccentric orbits, with the understandable exception of the tidally affected, short-period systems. The identification scheme that Duquennoy and Mayor had advanced in their monumental 1991 survey paper, a scheme fully in agreement with then-current theoretical models of star and planet formation, seemed to be well along the way to being proven to be correct.

Mayor's new objects also were remarkable for another reason. Marcy had noticed that the two brown dwarf companions HD 114762 B and 70 Virginis B had masses on the order of 10 Jupiter masses, far below the 80 Jupiter-mass cutoff for brown dwarf stars. Marcy then wondered why no one had yet found any brown dwarf companions with intermediate masses, say 30 Jupiter masses; after all, being much more massive, these companions should have been even easier to find than the ones found so far. Marcy therefore suggested that there was a "brown dwarf desert" in the range of masses between 10 and 80 or so Jupiter masses, where nothing existed at all. If it was true that such a major gap in mass existed, then the presumed "brown dwarf stars" HD 114762 B and 70 Virginis B might be better classified as "planets," or even as a new class of astronomical objects, neither planets nor brown dwarfs. It was important to know if Marcy's "brown dwarf desert" was real.

The newly announced objects did a good job of populating the "brown dwarf desert." In fact, Mayor's brown dwarfs looked much like a simple continuation to lower masses of normal stars, as they should if brown dwarfs were nothing more than severely mass-challenged stars. HD 114762 B and 70 Virginis B looked like the least massive brown dwarf stars, not planets or a new class of object.

While the discoveries were coming at a furious pace, at least it was easy so far to sort most of them into one basket or the other. But how long would that last? It would take only the discovery of a single, low-mass planet on an eccentric, long-period orbit to blow the entire identification scheme sky high. And who knew what would be found next?

JULY 3, 1996: The summer blockbuster movie *Independence Day* opened nationwide, with Washington crowds attending 4:00 A.M. shows for the experience of being the first to watch the White House get vaporized by beams from a comet-sized alien spacecraft. The possible down side of alien encounters became spectacularly apparent.

AUGUST 7, 1996: The case for life on other worlds took a giant and unexpected step forward when a team of scientists led by NASA's

David McKay and Everett Gibson announced that they had found interconnected lines of evidence for ancient microbial life on Mars. The evidence included images of bacteria-shaped tubules, complex organic compounds called polycyclic aromatic hydrocarbons, magnetic minerals commonly found inside terrestrial bacteria, and carbonate globules similar to those produced by bacteria. Taken separately, each piece of evidence was inconclusive, but McKay and Gibson argued that, taken together, the best explanation was *bacterial life on Mars*. While many scientists were cautious about the claim, if McKay and Gibson were right, we were not alone, even if our nearest known neighbors were as dumb as dirt when alive.

The Johnson Space Center scientists and their colleagues had been studying a rare meteorite, one of only a dozen ever found that are believed to have originated on Mars. The meteorite had the utilitarian name of ALH84001, because it was literally the first meteorite found in 1984 near the Allan Hills of Antarctica; the windswept Antarctic ice sheet is a prime location for finding new meteorites, which look very much out of place. ALH84001 had been dated at 4.5 billion years old, as old as Mars itself, and so this rock had been around from the beginning on Mars, including the first billion years or so, when liquid water flowed across the Martian landscape. McKay and Gibson believed that ALH84001 had gotten infected with tiny Martian bacteria that left behind fossilized evidence of their existence.

McKay and Gibson's article about their stunning discovery was about to be published in *Science*. They were trying to honor *Science*'s prohibition about press conferences before the August 16 publication date of the paper, but a reporter jumped the gun and broke the press embargo on August 6. NASA was deluged with inquiries and hastily organized a press conference for the next day. Dan Goldin had personally briefed President Bill Clinton, Vice-President Al Gore, and their close advisors about the impending announcement on July 30. As a result, then-presidential political consultant Dick Morris whispered on August 2 to call girl Sherry Rowlands that life had been found on Mars, but she misrecorded the thrilling secret in her diary as "life on Pluto."

The press went into a frenzy. For several days, the possibility of life on Mars was front page news on practically every newspaper in the country—*Time* and *Newsweek* put the story on their covers. A Brazilian journalist sought me out, while I was on vacation in Florida, and pleaded for an interview, even though my credentials for knowing about extraterrestrial life were pretty much limited to my mistaken appearance on the *America's Talking* television program; he might just as well have called humor columnist Dave Barry for an authoritative comment.

By chance, the day before McKay and Gibson's press conference, the National Research Council (NRC) had released a critique of NASA's plans for Mars exploration, including an admonishment to NASA to be sure that Martian rocks were collected and returned to Earth for study. ALH84001 had arrived on Earth all on its own, and there were un-doubtedly many more Martian meteorites waiting to be found on the Antarctic ice sheet, available practically for free compared to the cost of a round-trip ticket to Mars. Nevertheless, it was clear that if McKay and Gibson were right, NASA would have to mount a serious effort to look for life on Mars. The relevant NRC committee, COMPLEX, was called back into action to review once more NASA's plans for Mars, in partic-ular how to go about returning samples of the Martian surface.

President Bill Clinton was so impressed by the election-year discov-ery that, prodded by Maryland Senator Barbara Mikulski, he called on Vice-President Gore to hold a bipartisan "space summit" on the future of the U.S. space program. The sweet smell of money was in the air; both House Speaker Newt Gingrich and the Vice-President promised Goldin more funds for Mars exploration.

Centuries before, Herschel had used his hand-made telescope to study Mars, following the alternating appearance of the white polar caps during the Martian year. Herschel measured the inclination of Mars's rotational (spin) axis to its orbital plane and found it to be very similar to that of the Earth, about 23.5 degrees. Mars thus had seasons, just like the Earth, that explained the behavior of the polar caps. Com-bined with telescopic evidence for a thin atmosphere, Herschel con-cluded that Mars was the most similar to Earth of all the planets, and in 1784 he speculated that life might exist on Mars.

Over two hundred years later, McKay and Gibson had provided the best evidence yet for life on Mars. If bacterial life did once exist on Mars, that would be astounding enough, but if this putative life also *originated* on Mars, the implications would be profound. On the basis of a sample of two planets, Earth and Mars, the argument then could be made that primitive life inevitably formed wherever possible, wher-ever liquid water flowed over a planet with a solid surface and a stable orbit. Proof of the existence of Martian-bred life would make it likely that some sort of creatures lived on *any* Earth-like planets that hap-pened to orbit in the habitable zones of their stars.

AUGUST 15, 1996: Art decided not to imitate, but to ridicule life. Tele-vision commercials for the movie *Independence Day* made light of the Mars meteorite discovery by showing a sequence composed of the written statement "Last week scientists discovered evidence for life on other planets," followed by a large "DUH" and then by explosive visuals from the movie.

On the same day, the first allocations of NASA's share of the observing time on the Keck telescopes were made public. The lion's share of the first 27 nights available went to those who were planning to look for extrasolar planets and brown dwarfs: Cochran, Gatewood, Marcy, and Stauffer were among the winners who could expect to take a trip to the summit of Mauna Kea after October 1, 1996. Even though Mayor and Marcy had shown that Jupiter-mass extrasolar planets could be found with modest-sized telescopes, the Kecks offered the advantage of a huge light-collecting area, which meant that the Doppler shift of a nearby star could be determined in minutes instead of hours. That, in turn, meant that many more stars could be observed in a single night of observing, so that more planets could be caught and thrown into the bucket. Keck had instantly become the key instrument in the race to find more planets.

Curiously, Latham was not among the winners. Latham's spectroscopic approach to detecting extrasolar planets depended on using the Keck I telescope's HIRES spectrometer, but HIRES was of limited use without the iodine cell that provided the stable source of reference lines needed to determine the Doppler shift of the star. The iodine cell was listed as being available for all users when the first Keck Telescope Guide was issued on March 15, but that had turned out to be incorrect. The iodine cell had been built by Marcy, and it was considered to be a proprietary device, meaning that only Marcy and those designated by him (e.g., Cochran) could use it. This restriction was meant to be a suitable reward to Marcy for developing the iodine cell, but it had the effect of giving Marcy veto power over the decisions of the panel of scientists who ranked the proposals to use the Kecks. After their ill-fated encounter on ABC's *Nightline* program, Marcy was unwilling to allow Latham to use the iodine cell.

When the next Keck Telescope Guide was released in August 1996, the iodine cell for HIRES had been literally whited out of existence; nothing remained in its place in the Guide except for a blank space. NASA found itself having to make frantic plans to build a second iodine cell, if need be, that could be used by everyone, even Latham.

SEPTEMBER 19, 1996: The White House released a new National Space Policy that provided guidelines for NASA's future activities. While broad in most respects, the guidelines were remarkably specific and brief about what NASA should be doing in the area of space science.

Not surprisingly, coming as it did right after McKay and Gibson's bombshell, the Policy committed NASA to sending robotic explorers to the surface of Mars by the year 2000 and to gathering samples of Mars and the other bodies in the Solar System and returning them to the Earth for analysis. But that was not all. The Policy also committed NASA to "undertake a long-term program to identify and characterize planetary bodies in orbit around other stars."

The search for extrasolar planets had made it to the rarefied top of NASA's scientific priorities, at least as defined by the White House.

OCTOBER 23, 1996: The American Astronomical Society's Division for Planetary Science was holding its annual meeting in Tucson, and a press conference on an almost routine topic was about to be held. After eight years of searching, Bill Cochran was wearing a tweed jacket and his red hot chili peppers tie, ready to announce his first planetary detection. Cochran's discovery would turn out to be anything but routine, however.

Cochran and his Texas colleague Artie Hatzes had been measuring the Doppler shifts of two stars in a triple system called 16 Cygni ever since 1988, hoping to find the characteristic signal of an extrasolar Jupiter. The 16 Cygni system is about 70 light years away from the Earth, in the constellation Cygnus. The two stars Cochran and Hatzes were monitoring were 16 Cygni A and 16 Cygni B, two Sun-like stars that form a binary pair separated by a large enough distance (1,100 times the Earth-Sun distance) that planets should have plenty of elbow room to form around either star. The third member of 16 Cygni was a fainter, lower-mass star located more than twice as far away as the distance between 16 Cygni A and 16 Cygni B. If the third star was at least five times farther away, then 16 Cygni would be a hierarchical triple-star system, a configuration that should be stable for billions of years or more.

Cochran and Hatzes had begun to suspect by February 1996 that 16 Cygni B had a companion, but they wanted to collect more observations in order to be sure that they had a real detection. In the spirit of cooperation, Cochran and Marcy had exchanged information about the stars they were each observing that they thought were beginning to show signs of planets. Cochran told Marcy about 16 Cygni B, and Marcy told Cochran about a star called HR 7504. Several weeks passed before Marcy realized that HR 7504 was the *same star* as 16 Cygni B, which has several aliases, including the equally memorable SAO 31899 and HD 186408. They were both hot on the trail of the same planet. Marcy e-mailed Cochran about the coincidence, and they decided to wait until the Tucson meeting and then hold a joint press conference.

FIGURE 22. Artie Hatzes (left) and William Cochran (right) of the University of Texas's McDonald Observatory, codiscoverers (with Geoffrey Marcy and Paul Butler) of the first planet on an eccentric orbit, circling the solar-type star 16 Cygni B. The planet apparently owes its eccentricity to 16 Cygni B's binary companion. (Photographed by Marsha Miller. Used with permission, courtesy of Office of Public Affairs. Copyright 1996 by The University of Texas at Austin.)

Cochran announced in Tucson that the combined McDonald/Lick teams agreed that 16 Cygni B was orbited by a planet with a mass at least 50 percent greater than that of Jupiter. The orbital period was 2.2 years, meaning that the planet orbited at a distance about 70 percent greater than Earth's orbit. 16 Cygni B's planet was not another hot Jupiter but appeared to be a more normal gas giant planet, like Marcy and Butler's 47 Ursa Majoris B, orbiting well away from its star. 16 Cygni B's twin star, 16 Cygni A, showed no sign of having any extrasolar planets, however.

So far, so good. But while Cochran did not harp on the fact, there was something potentially disquieting about the new discovery. 16 Cygni B's companion had a mass that placed it well within the range of gas giant planets, but its orbit was *eccentric*. Not just a little eccentric, but a lot—an eccentricity of about 0.67, even more than that of the brown dwarfs HD 114762 B and 70 Virginis B. According to the identification scheme that had worked for every companion found so far, planets should be formed on roughly circular orbits, not on the highly elliptical orbit of 16 Cygni B's planet. Something was wrong here.

Marcy was extremely pleased about the discovery. He felt that 16 Cygni B's planet had totally invalidated the planet/brown dwarf identification scheme. If an object's orbital eccentricity could no longer be used to separate planets from brown dwarfs, he reasoned, then a far better case could be made, for example, that hefty 70 Virginis B was really an overweight gas giant planet and not a bulimic brown dwarf star. Marcy and Butler could then claim credit for having discovered yet another planet, rather than the least massive brown dwarf star. Marcy also concluded that because 16 Cygni B's planet had a noncircular orbit, our basic theory of planet formation, which had predicted formation on nearly circular orbits, had to be incorrect—we really did not understand how planets formed.

Marcy's teaching duties at San Francisco State University prevented him from attending the Tucson press conference, but he did send out an e-mail message to the community on the day of the press conference making his points about 70 Virginis B and the apparent failure of the theory of planet formation. Cochran, on the other hand, had put a different spin on the situation at the press conference, pointing out that there were two other stars in the 16 Cygni system and hinting that therein might lie the explanation for the oddly eccentric planet.

If 16 Cygni A had an orbit that ever brought it close to 16 Cygni B, its gravitational kicks would take a planet formed on an initially circular orbit and force its orbit to become eccentric. Alternatively, if the more distant third star, 16 Cygni C, had ever passed close to 16 Cygni B, the same effect would have occurred. Because the orbital periods of the 16 Cygni triple-star system were measured in many thousands of years, it would be a long time before astronomers were able to pinpoint the true orbital configuration of the 16 Cygni system, now and in the distant past. But if 16 Cygni B's planet had ended up on an eccentric orbit by virtue of its presence in a triple-star system, then Mayor's identification scheme could be salvaged, at least for planets around single stars, and the theory of planet formation could be pulled back out of the trashbin as well.

Like Cochran, Marcy, and Gatewood, Arizona's Bob McMillan had been searching for extrasolar planets since the late 1980s, but he alone still had no detections to show for his many years of effort. McMillan decided to cut his losses and move forward—he stopped taking data with the old system and began working with Gatewood on a plan to add a new spectrometer to Gatewood's astrometric planet finder for the Keck II telescope. Gatewood's sensitive electronic detectors did not need all the light gathered by the huge Keck mirrors, so McMillan would bleed off most of the excess starlight, which would otherwise be wasted, and use it to measure the star's Doppler velocity at the same time that Gatewood looked for its wobble. Two complementary planet searches could then be conducted at the same time, a wonderful idea indeed.

NOVEMBER 18, 1996: Matthew Holman and Scott Tremaine of the Canadian Institute for Theoretical Astrophysics and the University of Texas's Jihad Touma submitted a paper to *Nature* that advanced an explanation for the unexpectedly eccentric orbit of 16 Cygni B's gas giant planet. The paper showed that the planet could have formed on a nearly circular orbit and still gotten kicked into an elliptical orbit by the gravitational pull of 16 Cygni B's binary companion, 16 Cygni A. Their explanation rested on the guess that the giant planet's orbit around 16 Cygni B was inclined to the orbit of the two binary stars themselves (i.e., that all three bodies did not orbit in exactly the same plane). If this assumption was correct, Holman and his colleagues proved that the orbital eccentricity of the planet would increase and then decrease, over millions to billions of years. At any given time, there was a good chance that the planet would be found to have an eccentricity at least as high as 16 Cygni B's planet.

Holman's paper also showed that the gravitational effects of a binary star companion were unable to change the shapes of the orbits of planets that stayed very close to their stars (i.e., the hot Jupiters, like 55 Rho[1] Cancri B and Tau Bootis B). These planets simply stayed too deep within the clutch of their stars for another star to have any appreciable chance of changing their ways; 16 Cygni B's planet, on the other hand, was open to suggestions.

Mayor's planet versus brown dwarf identification scheme was saved, though a caveat would have to be added regarding planets like 16 Cygni B's. The theory of planet formation on circular orbits had survived a close call as well. If a planet on a highly elliptical orbit was ever found around a *single* star, though, theorists would have to ride to the rescue again.

DECEMBER 9–10, 1996: The SIM was in the early stages of planning by NASA, and that meant that a new Science Working Group had to be

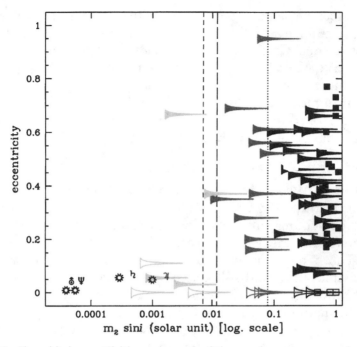

FIGURE 23. The orbital eccentricities and masses of the very-low-mass companions to solar-type stars discovered by the end of 1996. Brown dwarf stars lie just to the left of the vertical dotted line, which is the minimum mass for a hydrogen-burning star. The vertical dashed lines roughly separate brown dwarf stars from gas giant planets, the latter having nearly circular orbits, except for 16 Cygni B's planet, where 16 Cygni B's binary companion has been implicated. Open symbols represent companions whose primordial eccentricity has been altered by tidal interactions with the parent star. (Courtesy of Michel Mayor, Geneva Observatory.)

formed to participate in designing the mission—the SIMSWG. The SIMSWG held a meeting at the Space Telescope Science Institute in part to decide how capable SIM would have to be at detecting planets by the astrometric method.

While Mike Shao was the head of the interferometry effort at JPL, Shao shared the chair of SIMSWG with Deane Peterson, who had chaired the previous SISWG. Around the time of the NASA Ames meetings about extrasolar planet searches in 1976, Shao, Reasenberg, and Gatewood had agreed to keep each other involved in the planet search game, even if only one of them should succeed in winning the go-ahead to build a space-based telescope. Twenty years later, Shao had won decisively, and he graciously kept the decades-old promise—in spite of the many battles over OSI, POINTS, and AIM, both Reasenberg and Gatewood were made members of the SIMSWG. But regardless of the decades-old agreement, Reasenberg's and Gatewood's extremely specialized expertise and insight were simply too valuable to be wasted. In the end, the astrometrists stuck together, even if the spectroscopists did not.

The SIMSWG decided that SIM must be able to measure the wobble of stars to an accuracy of about 10 microarcseconds, a thousand times more accurate than van de Kamp's observations at Sproul. This level of accuracy would allow SIM to be able to detect planets with masses about 10 times that of Earth (i.e., planets similar to Uranus and Neptune in mass). When the Keck Interferometer was built, it was expected to be able to reach about the same level of accuracy, so it was clear that SIM had to do at least as well as the twin Kecks with their outriggers. While they could not promise anything, the JPL engineers would try to reach 1-microarcsecond accuracy, in order to provide a real improvement over the Keck Interferometer. SIM might then be able to find an Earth or two around nearby stars. But even if SIM did not find new Earths, it would help to find the nearby stars most likely to have Earths of their own.

SIM was expected to have a lifetime in space of only five years, which would severely restrict its ability to find planets on long-period orbits, like the outer planets of our Solar System. Of course, if Uranus-mass planets had migrated inward to shorter-period orbits, like the hot Jupiters evidently did, then SIM might find *them*. Once SIM proved its worth by finding a few new planets, there would undoubtedly be pressure on NASA to keep SIM working for another 5 or 10 years, in order to give SIM enough time to find long-period planets, too. If Uranus-mass planets could be found with long orbital periods, there would be a strong presumption that habitable Earth-mass planets might orbit there as well. We would then know where to point the Planet Finder infrared interferometer when it was launched after SIM.

DECEMBER 20, 1996: Carl Sagan died at the age of 62 after a several-year struggle with a rare bone marrow disease, a stunningly tragic finale to perhaps the most spectacular year ever for those interested in life on other worlds. When I first heard that Sagan had passed away, it was like the moment that I and most Americans experienced when we heard that President Kennedy had been assassinated, news shocking enough to be frozen in memory. More than anyone else, Sagan had spoken and written eloquently about the challenge and excitement of looking for extraterrestrial life, in our Solar System and beyond. His 1966 book with the Soviet astronomer Iosef Shklovskii, *Intelligent Life in the*

Universe, was a pioneering treatise that stimulated many to join in the search for extrasolar planets and life. In the end, Sagan lived just long enough to see his visionary dreams finally begin to turn into reality.

JANUARY 17, 1997:   The new year began with the death at the age of 90 of Clyde Tombaugh, the last person to find a new planet in the Solar System. Tombaugh had lived a satisfyingly long life, long enough to share fully in the joy of the discoveries that other solar systems truly do exist, that there was almost a certainty that uncounted Earths orbit around the myriad of stars visible in the night sky.

After his discovery of Pluto in 1930, Tombaugh was promptly awarded a medal by Britain's Royal Astronomical Society in 1931. Only then did he begin his undergraduate education in astronomy, in 1932 at the University of Kansas. Tombaugh was probably the only scientist to matriculate in college after having made one of the most important scientific discoveries of the century. It would have been nearly impossible to top his 1930 discovery of the last planet in our Solar System.

After a hiatus of 65 years, the exclusive club composed of astronomers who have found a new planet had gained seven new members by the end of 1996: Mayor, Queloz, Marcy, Butler, Gatewood, Cochran, and Hatzes. Undoubtedly, many more members will be inducted into this historic club in the years to come. But who will be the one that finds the first extrasolar *Earth*?

EPILOGUE

Whenever a new discovery is reported to the scientific world, they first say, "It is probably not true." Thereafter, when the truth of the new proposition has been demonstrated beyond question, they say, "Yes, it may be true, but it is not important." Finally, when sufficient time has elapsed fully to evidence its importance, they say, "Yes, surely it is important, but it is no longer new."

—Michel Eyquem Montaigne (1533–1592)

In mid-1995, there was not a single scientifically credible claim for an extrasolar planet, in spite of over 50 years of intensive searching. Two very good candidates for brown dwarf stars were known, but that was *all* that astronomers had to show for their Sisyphean labors.

A year and a half later, the long-fruitless field of extrasolar planet searches had produced an incredibly rich crop of gas giant planets and brown dwarf stars: depending on your tastes, four hot Jupiters, four normal Jupiters, and seventeen brown dwarf stars had been found. And the search had just begun in earnest—only a few hundred nearby stars had been given an initial degree of scrutiny and attention. Much more remained to be discovered in the coming years and decades.

The pace of discovery slowed considerably in 1997, with only a single new planet around a solar-type star being announced. Rho Coronae Borealis was found to have a companion with a mass of at least 1.1 Jupiter masses, moving on a circular orbit with a period of 40 days. The detection was accomplished by Harvard-Smithsonian's Robert Noyes and his team, using the same spectroscopic technique that had proven so successful.

Rho Coronae Borealis's planet, with an orbital radius of 0.23 times the Earth-Sun distance, was intriguing for two reasons. First, the circularity of its orbit had to be primordial, because the planet's orbit was much too far from its star to be affected by tidal forces. Rho Coronae

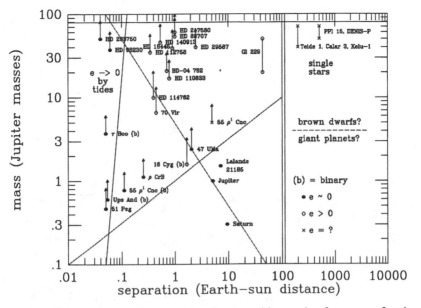

FIGURE 24. The search space for extrasolar planets and brown dwarf stars, as of early 1998, showing the plethora of discoveries, ranging from brown dwarf stars, to hot Jupiters, to Jupiters with orbits similar to Jupiter and Saturn. (Adapted from Alan Boss, 1996, *Physics Today*, volume 49, page 37.)

Borealis B thus obeyed the pattern exhibited so far by all the Jupiter-mass planets, consistent with Mayor's expectations, with the understandable exception of binary star 16 Cygni's planet. Second, Rho Coronae Borealis B had the largest orbital radius of any of the hot Jupiters, lying over twice as far from its star as 55 Rho[1] Cancri's planet and over four times farther than 51 Pegasi's and Upsilon Andromedae's. It was now obvious that hot Jupiters did not all reside at the same distance from their stars, suggesting that inward orbital migration of newly formed giant planets could be halted anywhere along the way. Giant planets might then be found in orbit anywhere between their star and their birthplace.

At the same time that the world was gaining a new planet, a serious effort was being made to eliminate one, the prototypical hot Jupiter itself, 51 Pegasi B. A rumor I had heard at the SIMSWG meeting the previous December soon surfaced in the form of a letter report in the February 27, 1997 issue of *Nature*. David Gray, an astronomer at the University of Western Ontario, had studied 51 Pegasi for many years, looking for evidence of a long-term cycle similar to the 11-year cycle that governs the Sun's magnetic fields and sunspots. After all the fuss about 51 Pegasi's planet arose, Gray decided to check his archive of observations. Gray reported that eight years of his data implied that 51 Pegasi was a *variable star*—oscillating in and out with a period of 4.23 days, exactly the same as that inferred for 51 Pegasi's planet!

If the surface of 51 Pegasi was really puffing up and down every 4.23 days, then the evidence for 51 Pegasi's giant planet would vanish in an instant. The velocity changes used by Mayor and Queloz to infer the presence of a planet would instead merely refer to the rise and fall of the star's surface. The other hot Jupiters might then be spurious, too, mere phantasms created by maliciously variable stars. The "new era" of the discovery of extrasolar planets would suddenly be tarnished, dragging us back to the painful lessons of Barnard's star and van Biesbroeck 8B. Gray's report exploded in the press with an intensity equal to that of the initial discovery.

Marcy and Queloz leapt to the defense of the imperiled planet, even before Gray's report was published, posting their reply on the World Wide Web. The question of the variability of 51 Pegasi had been raised by the referees of Mayor and Queloz's *Nature* paper, and there were good reasons why 51 Pegasi could not be a variable star. First, the brightness of variable stars changes as they swell and shrink in size, yet no hint of changes in 51 Pegasi's brightness had been found, in spite of highly sensitive searches. Second, Gray had found evidence for only a single period of oscillation of 51 Pegasi, whereas every other variable star found to date oscillates at many frequencies, like a struck bell. While troubling, Gray's data could not settle the issue.

One month later, Hatzes, Cochran, and a colleague published data in the *Astrophysical Journal* that showed no evidence of the effect found by Gray, though they admitted that 51 Pegasi's surface might still be oscillating at some level. At a conference in July 1997, Marcy presented a strong case in favor of the planetary interpretation of 51 Pegasi; Gray, who was in the audience, offered no rebuttal.

51 Pegasi became observable again in the summer of 1997, and everybody looked for the telltale signs of oscillations that could eliminate the first "confirmed" extrasolar planet. However, none of the four groups who searched could find any evidence for Gray's effect, not even Gray. Gray bravely surrendered in a letter to *Nature* in early 1998, admitting that his 1997 assertion that 51 Pegasi's planet did not exist was based on a combination of noisy data and bad luck.

51 Pegasi's planet is real. The seemingly endless cycle of discovery and disappointment has been broken forever.

The pulsar PSR1257+12 may have lost a planet in 1997, however. The evidence for the innermost planet, with the mass of the Moon and an orbital period of 25.3 days, was alleged to be nothing more than a spurious signal from our Sun, which rotates with the same period near its equator. The Sun expels a fierce wind of charged particles that rotates at the same rate as the Sun, and these charged particles might have created the 25.3 day variation in the pulsar timing data. Wolszczan planned further observations to settle the issue.

Gatewood's suspicion that a second planet lurked about the star Lalande 21185 fell prey to another year of astrometric observations. While the first planet (with a 32-year orbital period) continued to yank Lalande 21185 around, the second planet (with a 6-year period) refused to behave properly. Gatewood sagely pointed out that he now knew less with greater certainty.

In 1998, images were finally presented showing that Beta Pictoris was not the only normal star surrounded by a flattened disk of dust grains. The bright stars Fomalhaut and Vega, about 25 light years away, were found to have dust disks like Beta Pictoris. In addition, one of the stars in the binary system HR 4796 was observed to have a dust disk. The presence of lithium in the companion star implied a young age for the system, about 10 million years, an age when planets would be in the final phases of their formation.

The Keck Observatory will discover a stream of new extrasolar planets for years to come. Marcy's and Cochran's spectroscopic searches with the Keck I telescope will begin to bear fruit, as more stars with hot Jupiters and even hot Saturns are likely to be discovered. New planet searches with the newly nonproprietary iodine cell on HIRES were launched in 1997—Latham finally got his Keck telescope time and received friendly advice from Marcy on the best use of HIRES. Gatewood's astrometric device began measuring the precise locations of nearby stars, measurements that will slowly reveal the presence of Jupiter-sized and even much smaller planets on long-period orbits. Prolonged spectroscopic studies will definitively reveal the presence or absence of other planets in the systems already discovered: The mystery of 55 Rho[1] Cancri B's putative sibling will be solved. New multiple-planet systems will be found, and their characteristics will give us long-sought clues about how planet formation and subsequent orbital evolution proceeds around other stars.

Mayor began observations in June 1998 with a new spectrometer mounted on a modest-sized telescope at the European Southern Observatory in Chile. By starting to search in the southern hemisphere,

Mayor planned to have much of the night sky all to himself—the Keck telescopes will not be able to see many of the stars that Mayor will be able to monitor. However, Marcy and Butler were unwilling to cede the southern hemisphere to Mayor and Queloz. Marcy and Butler began their own radial velocity survey of the southern skies in September 1997, using an Australian telescope.

Weiler got everything he asked for when the President's budget for the years 1998 through 2002 was unveiled in early February 1997. Weiler's newly created Origins theme was promised the funds that would be necessary to build the Keck Interferometer, SIM, and *HST*'s successor, NGST. NASA's Martian exploration program was to be accelerated, with the first return of Martian samples scheduled for 2005 instead of 2007.

There was no need to hold a Presidential Space Summit to build support for NASA's Space Science initiatives—the money was already in the bag. After the bloodbath at Woods Hole in 1991, NASA's 1997 Strategic Planning meeting was moved to Breckenridge, Colorado, where Huntress presented a strawman mission plan, defined to an unprecedented degree by the winners in the President's budget, that won immediate acceptance.

The Breckenridge love-in made it clear, however, that NGST would be launched before the Planet Finder telescope. Huntress's strawman plan at Breckenridge placed NGST ahead of the Planet Finder. In an NRC report presented to Huntress on April 8, 1997, a committee of distinguished astronomers had given high priority to the search for extrasolar planets but advised that looking for terrestrial planets with the Planet Finder would be "premature" because such a search should be the "culmination, not the beginning" of the process. Rather than find extrasolar terrestrial planets prematurely, an epochal event anticipated for centuries if not millenia, this group of astronomers preferred to ensure that *HST* would have a prompt successor in the guise of NGST.

On Independence Day, July 4, 1997, NASA's *Pathfinder* spacecraft bounced to a scheduled landing on the surface of Mars, beginning a new era of sustained Mars exploration, with a primary goal being the search for life on the most Earth-like planet in the Solar System. The popular press reacted enthusiastically to NASA's return to the Red Planet, with daily front page updates on the progress of *Pathfinder* and its tiny robotic explorer, *Sojourner*. Vice-President Gore mounted a panoramic view of the Martian landscape in his office and invited visitors to don special glasses and explore Mars in 3D. Around the year 2005, NASA plans to return a promising sample of the Martian surface to Earth, so that terrestrial laboratories can perform the same sort of sophisticated search for signs of life that led to the dramatic announcement about the Martian meteorite ALH84001.

Sadly, Jürgen Rahe did not live to witness NASA's triumphant return to the Martian surface. Rahe was killed by a falling tree during a violent thunderstorm near his home in Potomac, Maryland on June 18, 1997. Rahe had led NASA's Solar System Exploration Division not only in planning the return to Mars, but also in laying the groundwork for NASA's search for other planetary systems.

The scientific controversy over the significance of ALH84001 became increasingly vitriolic in 1997, with the merits and failings of the evidence for Martian life being vigorously debated at conferences, in scientific journals, and in the press. It became clear that the evidence for Martian life was so complicated and convoluted that a quick, clear resolution of the question would not be forthcoming anytime soon. NASA and NSF did their best to ensure an eventual answer by funding new research programs devoted almost solely to the question of Martian microbes.

In early 1998, back-to-back papers published in *Science* showed that the bulk of the organic matter in ALH84001 is not Martian but terrestrial in origin, produced by the contaminating effects of 13,000 years of soaking in Antarctic ice meltwater. Both studies, however, admitted that there could be a small nonterrestrial organic component in ALH84001, giving McKay and Gibson enough maneuvering room to maintain their point of view about life on Mars. The Martian life debate continues.

Sometime after the year 2000, the NSF hopes to have built the Millimeter Array (MMA), a ring of 40 millimeter-wavelength antennas each 8 meters in diameter. A high-altitude desert site in Chile offers superb visibility through the Earth's atmosphere and enough room to spread out the antennas to form an interferometer miles wide. The MMA would be an impressive telescope for probing deep within the disks of gas and dust where planetary systems form.

Around the same time, giant visible-wavelength interferometers will be constructed, at both the Keck Observatory and at the European Southern Observatory. These interferometers may well have the astrometric ability to discover Neptune-mass planets around nearby stars.

In early 1997, NASA hoped to launch the SIM by the year 2005. SIM will perform a variety of ultraprecise astrometric measurements, including the search for extrasolar Neptunes and perhaps Earths, and will provide an important test of much of the interferometer technology necessary for the next step in space. If extrasolar Neptunes are found, then a major piece of the planet formation puzzle will have been

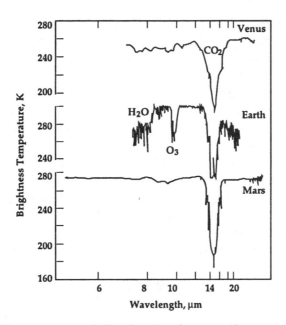

FIGURE 25. Earth's spectrum reveals the signature of an atmosphere containing carbon dioxide, water, and ozone, three key ingredients for a habitable planet. The goal of the Planet Finder telescope is to search for similar evidence for habitable Earth-like planets. (Courtesy of Charles Beichman, JPL. Adapted from *The Infrared Handbook*, editors W. L. Wolfe and G. J. Zissis, 1985, a report of the Environmental Research Institute of Michigan.)

found—Neptune-mass planets are thought to be the seeds for giant planet formation. Furthermore, because Neptunes and Earths are both thought to form from collisions between planetesimals, if extrasolar Neptunes exist, then almost certainly so do extrasolar Earths.

If all goes well with SIM, around the year 2010 NASA plans to launch the Planet Finder, a visionary space infrared interferometer that will be able to image nearly *all* the planets orbiting around nearby stars, Jupiters, Neptunes, and Earths alike. The Planet Finder will produce photographs of dots of light close to these stars, dots that with time will move around and around their stars, silently obeying the laws of planetary motion propounded 400 years ago by Kepler. Some of those dots will have the properties of an Earth-like planet, and some will orbit their star at a distance such that oceans and lakes could exist on their rocky surfaces. Those dots will be subjected to the piercing stare of the Planet Finder, which will capture enough light from the planets to begin to look for the atmospheric signatures of an inhabitable Earth—carbon dioxide, water, and ozone, a form of oxygen. If methane is found as well, that will be taken as excellent evidence that not only is the planet inhabitable, it is *inhabited*—methane could not last long in the presence of oxygen without a biological source. *We will have found another Earth.*

That discovery will be agonizing. We will know about the existence of another Earth—seemingly so close, yet so far beyond our reach. We may then decide to build a truly gigantic telescope in space, in order to begin to image the surface of this new Earth, at least in crude terms. We might hope to see evidence of clouds, oceans, ice caps, and continents, as Dan Goldin dreamed in 1994. Such details would be extraordinarily hard to obtain, but even these images would not completely slake our curiosity about the planet—we would only want to know more and more.

Looking farther into the future, once we discover nearby Earth-like planets that appear to support life, we undoubtedly will want to journey to these new worlds, perhaps first by launching miniature robotic spacecraft at high speeds. The interstellar spy satellites will image the planets as they shoot past and then transmit their precious information back to us at the speed of light. The space probes themselves would only be able to travel at speeds much less than that of light, so that it might take hundreds of years or more for the probes to reach their target Earths, a maddeningly long but necessary interval.

The images and information gathered may be so tantalizing that eventually our distant descendants will decide to launch spacecraft capable of colonizing the new worlds—science fiction will become science reality.

The scientific justification for the search for extrasolar Earths is absolutely compelling—no other area of space science can claim to have a goal as exciting, understandable, and promising as the search for new planetary systems. A new frontier in space has been opened by the events of 1995–1996, and human beings will find the challenge to be irresistible. The fiery urge to explore new worlds will inevitably be rekindled on an interstellar scale.

In the distant future, a thousand or two years from now, aliens will reach an Earth-like planet orbiting a star in the Sun's neighborhood. Perhaps the star will be named Lalande 21185, or maybe Tau Ceti or Epsilon Eridani. Our descendants will be those aliens.

LIST OF ACRONYMS

AIM—Astrometric Interferometry Mission

ASEPS—Astronomical Studies of Extrasolar Planetary Systems

ATF—Astrometric Telescope Facility

Caltech—California Institute of Technology

CDC—Control Data Corporation

CIT—Circumstellar Imaging Telescope

CIW—Carnegie Institution of Washington

COMPLEX—Committee on Planetary and Lunar Exploration

CRAF—Comet Rendevous/Asteroid Flyby

DCL—"Dear Colleague" Letter

DEC—Digital Equipment Corporation

DTM—Department of Terrestrial Magnetism

ESA—European Space Agency

ESO—European Southern Observatory

ExNPS—Exploration of Neighboring Planetary Systems

FRESIP—Frequency of Earth-sized Inner Planets

GPS—Global Positioning System

HIRES—High Resolution Echelle Spectrometer

HRMS—High Resolution Microwave Survey

HST—Hubble Space Telescope

HST&B—Hubble Space Telescope and Beyond

IAU—International Astronomical Union

IBM—International Business Machines

IRAS—Infrared Astronomical Satellite

IRTF—Infrared Telescope Facility

JPL—Jet Propulsion Laboratory

LGM—Little Green Men

MMA—Millimeter Array

MMT—Multiple Mirror Telescope (later, Monolithic Mirror Telescope)

NASA—National Aeronautics and Space Administration

NGST—Next Generation Space Telescope

NRC—National Research Council

NSF—National Science Foundation

OSI—Orbiting Stellar Interferometer

POINTS—Precision Optical Interferometer in Space

PSSWG—Planetary Systems Science Working Group

SETI—Search for Extraterrestrial Intelligence

SIM—Space Interferometry Mission

SIMSWG—Space Interferometry Mission Science Working Group

SISWG—Space Interferometry Science Working Group

SWG—Science Working Group

TOPS—Toward Other Planetary Systems

TOPSSWG—Toward Other Planetary Systems Science Working Group

TRW—Thompson-Ramo-Wooldridge Corporation

UC—University of California

UCLA—University of California, Los Angeles

UCSB—University of California, Santa Barbara

UCSC—University of California, Santa Cruz

UCSD—University of California, San Diego

USF—University of South Florida

VLT—Very Large Telescope

VLTI—Very Large Telescope Interferometer

SELECTED BIBLIOGRAPHY

Abell, G. *Exploration of the Universe.* New York: Holt, Rinehart, and Winston, 1964.

Abetti, G. *The History of Astronomy.* London: Sidgwick and Jackson, 1954.

Angel, J. R. P., and A. Burrows. "Seeking Planets Around Nearby Stars," *Nature* 374 (1995): 678.

Angel, J. R. P., A. Y. S. Cheng, and N. J. Woolf. "A Space Telescope for Infrared Spectroscopy of Earth-like Planets," *Nature* 322 (1986): 341.

Angel, J. R. P., and N. J. Woolf. "Searching for Life on Other Planets," *Scientific American,* April 1996, 60.

Angel, J. R. P., and N. J. Woolf. "An Imaging Nulling Interferometer to Study Extrasolar Planets," *Astrophysical Journal* 475 (1997): 373.

Artymowicz, P., C. J. Clarke, S. H. Lubow, and J. E. Pringle. "The Effect of an External Disk on the Orbital Elements of a Central Binary," *Astrophysical Journal* 370 (1991): L35.

Aumann, H. H., F. C. Gillett, C. A. Beichman, T. DeJong, J. R. Houck, F. J. Low, G. Neugebauer, R. G. Walker, and P. R. Wesselius. "Discovery of a Shell around Alpha Lyrae," *Astrophysical Journal* 278 (1984): L23.

Backer, D., S. Sallmen, and R. Foster. "Pulsar's Double Period Confirmed," *Nature* 358 (1992): 24.

Bailes, M., A. G. Lyne, and S. L. Shemar. "A Planet Orbiting the Neutron Star PSR1829-10," *Nature* 352 (1991): 311.

Basri, G., G. W. Marcy, and J. R. Graham. "Lithium in Brown Dwarf Candidates: The Mass and Age of the Faintest Pleiades Stars," *Astrophysical Journal* 458 (1996): 600.

Beckwith, S. V. W., and A. I. Sargent. "Circumstellar Disks and the Search for Neighbouring Planetary Systems," *Nature* 383 (1996): 139.

Beichman, C. A., ed. *A Road Map for the Exploration of Neighboring Planetary Systems.* Pasadena: NASA Jet Propulsion Laboratory, 1996.

Beichman, C. A., P. C. Myers, J. P. Emerson, S. Harris, R. Mathieu, P. J. Benson, and R. E. Jennings. "Candidate Solar-type Protostars in Nearby Molecular Cloud Cores," *Astrophysical Journal* 307 (1986): 337.

Berry, A. *A Short History of Astronomy.* New York: Dover, 1898.

Black, D. C., ed. *Project Orion: A Design Study of a System for Detecting Extrasolar Planets.* Washington, D.C.: Government Printing Office, 1980.

Black, D. C. "It's All in the Timing," *Nature* 352 (1991): 278.

Black, D. C. "Completing the Copernican Revolution: The Search for Other Planetary Systems," *Annual Review of Astronomy and Astrophysics* 33 (1995): 359.

Black, D. C., and M. S. Matthews, eds. *Protostars & Planets II.* Tucson: University of Arizona Press, 1985.

Black, D. C., and G. C. J. Suffolk. "Concerning the Planetary System of Barnard's Star," *Icarus* 19 (1973): 353.

Bodenheimer, P., and J. B. Pollack. "Calculations of the Accretion and Evolution of Giant Planets: The Effects of Solid Cores," *Icarus* 67 (1986): 391.

Bonnell, I., H. Martel, P. Bastien, J.-P. Arcoragi, and W. Benz. "Fragmentation of Elongated Cylindrical Clouds. III. Formation of Binary and Multiple Systems," *Astrophysical Journal* 377 (1991): 553.

Boss, A. P. "Protostellar Formation in Rotating Interstellar Clouds. VII. Opacity and Fragmentation," *Astrophysical Journal* 331 (1988): 370.

Boss, A. P. "Formation of Binary Stars," in *The Realm of Interacting Binary Stars,* eds. J. Sahade, G. McCluskey, and Y. Kondo. Dordrecht, The Netherlands: Kluwer, 1993.

Boss, A. P. "Evolution of the Solar Nebula. II. Thermal Structure during Nebula Formation," *Astrophysical Journal* 417 (1993): 351.

Boss, A. P. "Proximity of Jupiter-like Planets to Low Mass Stars," *Science* 267 (1995): 360.

Boss, A. P. "Giants and Dwarfs Meet in the Middle," *Nature* 379 (1996): 397.

Boss, A. P. "Extrasolar Planets," *Physics Today* 49 (1996): 32.

Boss, A. P., and P. Bodenheimer. "Fragmentation in a Rotating Protostar: A Comparison of Two 3-D Computer Codes," *Astrophysical Journal* 234 (1979): 289.

Bracewell, R. N. "Detecting Nonsolar Planets by Spinning Infrared Interferometer," *Nature* 274 (1978): 780.

Bracewell, R. N., and R. H. MacPhie. "Searching for Nonsolar Planets," *Icarus* 38 (1979): 136.

Brown, R. A., and C. J. Burrows. "On the Feasibility of Detecting Extrasolar Planets by Reflected Starlight Using the Hubble Space Telescope," *Icarus* 87 (1990): 484.

Brush, S. G. "From Bump to Clump: Theories of the Origin of the Solar System 1900–1960," in *Space Science Comes of Age,* eds. P. A. Hanle and V. D. Chamberlain. Washington, D.C.: Smithsonian Institution Press, 1981.

Brush, S. G. "Theories of the Origin of the Solar System 1956–1985," *Reviews of Modern Physics* 62 (1990): 43.

Burke, B. F. "Detection of Planetary Systems and the Search for Evidence of Life," *Nature* 322 (1986): 340.

Burke, B. F., ed. *TOPS: Toward Other Planetary Systems.* Washington, D.C.: NASA Solar System Exploration, 1992.

Burke, B. F., J. H. Rahe, and E. E. Roettger, eds. *Planetary Systems: Formation, Evolution, and Detection.* Dordrecht, The Netherlands: Kluwer, 1994.

Burrows, A., D. Saumon, T. Guillot, W. B. Hubbard, and J. I. Lunine. "Prospects for Detection of Extra-solar Giant Planets by Next-generation Telescopes," *Nature* 375 (1995): 299.

Burrows, C. J., K. R. Stapelfeldt, A. M. Watson, J. E. Krist, G. E. Ballester, J. T. Clarke, D. Crisp, J. S. Gallagher, R. E. Griffiths, J. G. Hoessel, J. A. Holtzman, J. R. Mould, P. A. Scowen, J. T. Trauger, and J. A. Westphal. "Hubble Space Telescope Observations of the Disk and Jet of HH 30," *Astrophysical Journal* 473 (1996): 437.

Butler, R. P., and G. W. Marcy. "A Planet Orbiting 47 Ursae Majoris," *Astrophysical Journal* 464 (1996): L153.

Butler, R. P., G. W. Marcy, E. Williams, C. McCarthy, P. Dosanjh, and S. S. Vogt. "Attaining Doppler Precision of 3 m/s," *Publications of the Astronomical Society of the Pacific* 108 (1996): 500.

Butler, R. P., G. W. Marcy, E. Williams, H. Hauser, and P. Shirts. "Three New '51 Pegasi-Type' Planets," *Astrophysical Journal* 474 (1997): L115.

Cameron, A. G. W. "Accumulation Processes in the Primitive Solar Nebula," *Icarus* 18 (1973): 407.

Cameron, A. G. W. "Physics of the Primitive Solar Accretion Disk," *Moon & Planets* 18 (1978): 5.

Campbell, B., and G. A. H. Walker. "Precision Radial Velocities with an Absorption Cell," *Publications of the Astronomical Society of the Pacific* 91 (1979): 540.

Campbell, B., G. A. H. Walker, and S. Yang. "A Search for Substellar Companions to Solar-type Stars," *Astrophysical Journal* 331 (1988): 902.

Cochran, W. D., and A. P. Hatzes. "Radial Velocity Searches for Other Planetary Systems," *Astrophysics and Space Science* 241 (1996): 43.

Cochran, W. D., A. P. Hatzes, R. P. Butler, and G. W. Marcy. "The Discovery of a Planetary Companion to 16 Cygni B," *Astrophysical Journal* 483 (1997): 457.

Cochran, W. D., A. P. Hatzes, and T. J. Hancock. "Constraints on the Companion Object to HD 114762," *Astrophysical Journal* 380 (1991): L35.

Coppenbarger, D. S., T. J. Henry, and D. W. McCarthy. "Ross 614AB: A Redetermination of the Masses One Orbit Later," *Astronomical Journal* 107 (1994): 1551.

Croswell, K. "Does Barnard's Star Have Planets?" *Astronomy* 16 (1988): 6.

Croswell, K. "The Story of Lalande 21185," *Sky & Telescope,* June 1995, 68.

Duquennoy, A., and M. Mayor. "Multiplicity among Solar-type Stars in the Solar Neighborhood II. Distribution of the Orbital Elements in an Unbiased Sample," *Astronomy & Astrophysics* 248 (1991): 485.

Edgeworth, K. E. "The Origin and Evolution of the Solar System," *Monthly Notices of the Royal Astronomical Society* 109 (1949): 600.

Forrest, W. J., M. F. Skrutskie, and M. Shure. "A Possible Brown Dwarf Companion to Gliese 569," *Astrophysical Journal* 330 (1988): L119.

Gamow, G., and J. A. Hynek. "A New Theory by C. F. von Weizsacker of the Origin of the Planetary System," *Astrophysical Journal* 101 (1945): 249.

Gatewood, G. "An Astrometric Study of Lalande 21185," *Astronomical Journal* 79 (1974): 52.

Gatewood, G. "On the Astrometric Detection of Neighboring Planetary Systems," *Icarus* 27 (1976): 1.

Gatewood, G. D. "A Study of the Astrometric Motion of Barnard's Star," *Astrophysics and Space Science* 223 (1995): 91.

Gatewood, G. W. "Lalande 21185," *Bulletin of the American Astronomical Society* 28 (1996): 885.

Gatewood, G., and H. Eichhorn. "An Unsuccessful Search for a Planetary Companion of Barnard's Star," *Astronomical Journal* 78 (1973): 769.

Gehrels, T., ed. *Protostars & Planets*. Tucson: University of Arizona Press, 1978.

Goldreich, P., and S. Tremaine. "Disk-Satellite Interactions," *Astrophysical Journal* 241 (1980): 425.

Goldreich, P., and W. R. Ward. "The Formation of Planetesimals," *Astrophysical Journal* 183 (1973): 1051.

Grady, M., I. Wright, and C. Pillinger. "Opening a Martian Can of Worms?" *Nature* 382 (1996): 575.

Graham, J. R., K. Matthews, G. Neugebauer, and B. T. Soifer. "The Infrared Excess of G29-38: A Brown Dwarf or Dust?" *Astrophysical Journal* 357 (1990): 216.

Gray, D. F. "Absence of a Planetary Signature in the Spectra of the Star 51 Pegasi," *Nature* 385 (1997): 795.

Gray, D. F. "A Planetary Companion for 51 Pegasi Implied by Absence of Pulsations in the Stellar Spectra," *Nature* 391 (1998): 153.

Gray, D. F., and A. P. Hatzes. "Non-Radial Oscillation in the Solar-Temperature Star 51 Pegasi," *Astrophysical Journal* 490 (1997): 412.

Guillot, T., A. Burrows, W. B. Hubbard, J. I. Lunine, and D. Saumon. "Giant Planets at Small Orbital Distances," *Astrophysical Journal* 459 (1996): L35.

Haas, M., and C. Leinert. "Search for the Suspected Brown Dwarf Companion to Giclas 29-38 Using IR-slit-scans," *Astronomy & Astrophysics* 230 (1990): 87.

Harrington, R. S., and B. J. Harrington. "Barnard's Star: A Status Report on an Intriguing Neighbor," *Mercury*, May–June 1987, 77.

Harrington, R. S., V. V. Kallarakal, and C. C. Dahn. "Astrometry of the Low-luminosity Stars VB8 and VB10," *Astronomical Journal* 88 (1983): 1038.

Hatzes, A. P., W. D. Cochran, and E. J. Bakker. "Further Evidence for the Planet around 51 Pegasi," *Nature* 391 (1998): 154.

Hatzes, A. P., W. D. Cochran, and C. M. Johns-Krull. "Testing the Planet Hypothesis: A Search for Variability in the Spectral-Line Shapes of 51 Pegasi," *Astrophysical Journal* 478 (1997): 374.

Heintz, W. D. "Astrometric Study of Four Binary Stars," *Astronomical Journal* 77 (1972): 160.

Heintz, W. D. "The Substellar Masses of Wolf 424," *Astronomy & Astrophysics* 217 (1989): 145.

Heintz, W. D. "The Substellar Masses of Wolf 424. II," *Astronomy & Astrophysics* 277 (1993): 452.

Heintz, W. D. "Photographic Astrometry of Binary and Proper-motion Stars. VIII," *Astronomical Journal* 108 (1994): 2338.

Henry, G. W., S. L. Baliunas, R. A. Donahue, W. H. Soon, and S. H. Saar. "Properties of Sun-Like Stars with Planets: 51 Pegasi, 47 Ursae Majoris, 70 Virginis, and HD 114762," *Astrophysical Journal* 474 (1997): 503.

Henry, T. J., D. S. Johnson, D. W. McCarthy, and J. D. Kirkpatrick. "Red/Infrared Observations of Wolf 424 AB: Are the Components Substellar?" *Astronomy & Astrophysics* 254 (1992): 116.

Hershey, J. L. "Astrometric Analysis of the Field of AC+65," *Astronomical Journal* 78 (1973): 421.

Hills, J. G. "Planetary Companions of Pulsars," *Nature* 226 (1970): 730.

Hoffleit, D. "New Unseen Companions," *Sky & Telescope,* June 1944, 14.

Holman, M., J. Touma, and S. Tremaine. "Chaotic Variations in the Eccentricity of the Planet Orbiting 16 Cygni B," *Nature* 386 (1997): 254.

Hoyle, F. "On the Fragmentation of Gas Clouds into Galaxies and Stars," *Astrophysical Journal* 118 (1953): 513.

Hunter, A. "Non-Solar Planets," *Nature* 152 (1943): 66.

Isaacman, R., and C. Sagan. "Computer Simulations of Planetary Accretion Dynamics: Sensitivity to Initial Conditions," *Icarus* 31 (1977): 510.

Jeans, J. H. "Origin of the Solar System," *Nature* 149 (1942): 695.

Jensen, O. G., and T. Ulrych. "An Analysis of the Perturbations on Barnard's Star," *Astronomical Journal* 78 (1973): 1104.

Kafatos, M. C., R. S. Harrington, and S. P. Maran, eds. *Astrophysics of Brown Dwarfs.* Cambridge, England: Cambridge University Press, 1986.

Kasting, J. F., D. P. Whitmire, and R. T. Reynolds. "Habitable Zones around Main Sequence Stars," *Icarus* 101 (1993): 108.

Kuiper, G. P. "On the Origin of the Solar System," *Proceedings of the National Academy of Science* 37 (1951): 1.

Kumar, S. S. "Planetary Systems," in *The Emerging Universe,* eds. W. C. Saslaw and K. C. Jacobs. Charlottesville: University Press of Virginia, 1972.

Latham, D. W., T. Mazeh, R. P. Stefanik, M. Mayor, and G. Burki. "The Unseen Companion of HD114762: A Probable Brown Dwarf," *Nature* 339 (1989): 38.

Leger, A., J. M. Mariotti, B. Mennesson, M. Ollivier, J. L. Puget, D. Rouan, and J. Schneider. "Could We Search for Primitive Life on Extrasolar Planets in the Near Future? The DARWIN Project," *Icarus* 123 (1996): 249.

Levin, A. E. "The Otto Schmidt School and the Development of Planetary Cosmogony in the USSR," in *The Origin of the Solar System: Soviet Research 1925–1991,* eds. A. E. Levin and S. G. Brush. New York: American Institute of Physics, 1995.

Levy, E. H., and J. I. Lunine, eds. *Protostars and Planets III.* Tucson: University of Arizona Press, 1993.

Lin, D. N. C., P. Bodenheimer, and D. C. Richardson. "Orbital Migration of the Planetary Companion of 51 Pegasi to its Present Location," *Nature* 380 (1996): 606.

Lin, D. N. C., and J. Papaloizou. "On the Structure and Evolution of the Primordial Solar Nebula," *Monthly Notices of the Royal Astronomical Society* 191 (1980): 37.

Lissauer, J. J. "Timescales for Planetary Accretion and the Structure of the Protoplanetary Disk," *Icarus* 69 (1987): 249.

Low, C., and D. Lynden-Bell. "The Minimum Jeans Mass or When Fragmentation Must Stop," *Monthly Notices of the Royal Astronomical Society* 176 (1976): 367.

Lunine, J. I., W. B. Hubbard, and M. S. Marley. "Evolution and Infrared Spectra of Brown Dwarfs," *Astrophysical Journal* 310 (1986): 238.

Lynden-Bell, D., and J. E. Pringle. "The Evolution of Viscous Disks and the Origin of the Nebular Variables," *Monthly Notices of the Royal Astronomical Society* 168 (1974): 603.

Mackay, A. L. *A Dictionary of Scientific Quotations.* Bristol, England: Institute of Physics, 1991.

Malhotra, R., D. Black, A. Eck, and A. Jackson. "Resonant Orbital Evolution in the Putative Planetary System of PSR1257+12," *Nature* 356 (1992): 583.

Marcy, G. W. "Back in Focus," *Nature* 391 (1998): 127.

Marcy, G. W., and K. J. Benitz. "A Search for Substellar Companions to Low-mass Stars," *Astrophysical Journal* 344 (1989): 441.

Marcy, G. W., and R. P. Butler. "A Planetary Companion to 70 Virginis," *Astrophysical Journal* 464 (1996): L147.

Marcy, G. W., R. P. Butler, E. Williams, L. Bildsten, J. R. Graham, A. M. Ghez, and J. G. Jernigan. "The Planet around 51 Pegasi," *Astrophysical Journal* 481 (1997): 926.

Martin, E. L., R. Rebolo, and M. R. Zapatero-Osorio. "Spectroscopy of New Substellar Candidates in the Pleiades: Toward a Spectral Sequence for Young Brown Dwarfs," *Astrophysical Journal* 469 (1996): 706.

Mayor, M., and D. Queloz. "A Jupiter-mass Companion to a Solar-type Star," *Nature* 378 (1995): 355.

Mayor, M., and D. Queloz. "A Search for Substellar Companions to Solar-type Stars via Precise Doppler Measurements: A First Jupiter Mass Companion Detected," in *Cool Stars, Stellar Systems, and the Sun, 9th Cambridge Workshop,* eds. R. Pallavicini and A. K. Dupree. San Francisco: Astronomical Society of the Pacific, 1996.

Mayor, M., D. Queloz, S. Udry, and J. L. Halbwachs. "From Brown Dwarfs to Planets," in *Astronomical and Biochemical Origins and Search for Life in the Universe,* eds. C. B. Cosmovici, S. Bowyer, and D. Werthimer. Bologna: Editrice Compositori, 1997.

Mazeh, T., Y. Krymolowski, and G. Rosenfeld. "The High Eccentricity of the Planet Orbiting 16 Cygni B," *Astrophysical Journal* 477 (1997): L103.

McCarthy, D. W., R. G. Probst, and F. J. Low. "Infrared Detection of a Close Cool Companion to Van Biesbroeck 8," *Astrophysical Journal* 290 (1985): L9.

McCaughrean, M. J., and C. R. O'Dell. "Direct Imaging of Circumstellar Disks in the Orion Nebula," *Astronomical Journal* 111 (1996): 1977.

McKay, D. S., E. S. Gibson, K. L. Thomas-Kleprta, H. Vali, C. S. Romanek, S. J. Clemett, and D. F. Chillier. "Search for Past Life on Mars: Possible Relic Biogenic Activity in Martian Meteorite ALH84001," *Science* 273 (1996): 924.

Mizuno, H. "Formation of the Giant Planets," *Progress of Theoretical Physics* 64 (1980): 544.

Moore, P. "The Hunt for Neptune," *Sky & Telescope,* September 1996, 42.

Mumford, G. S. "The Legacy of E. E. Barnard," *Sky & Telescope,* July 1987, 30.

Nakajima, T. "Planet Detectability by an Adaptive Optics Stellar Coronagraph," *Astrophysical Journal* 425 (1994): 348.

Nakajima, T., B. R. Oppenheimer, S. R. Kulkarni, D. A. Golimowski, K. Matthews, and S. T. Durrance. "Discovery of a Cool Brown Dwarf," *Nature* 378 (1995): 463.

Noyes, R. W., S. Jha, S. G. Korzennik, M. Krockenberger, P. Nisenson, T. M. Brown, E. J. Kennelly, and S. D. Horner. "A Planet Orbiting Around the Star Rho Coronae Borealis," *Astrophysical Journal* 483 (1997): L111.

Oppenheimer, B. R., S. R. Kulkarni, K. Matthews, and T. Nakajima. "Infrared Spectrum of the Cool Brown Dwarf Gl 229B," *Science* 270 (1995): 1478.

Pasachoff, J. M. *Astronomy: From the Earth to the Universe,* Philadelphia: Saunders, 1991.

Peale, S. J. "On the Verification of the Planetary System around PSR1257+12," *Astronomical Journal,* 105 (1993): 1562.

Perrier, C., and J. M. Mariotti. "On the Binary Nature of Van Biesbroeck 8," *Astrophysical Journal* 312 (1987): L27.

Podsiadlowski, P., J. E. Pringle, and M. J. Rees. "The Origin of the Planet Orbiting PSR1829-10," *Nature* 352 (1991): 783.

Pollack, J. B. "Origin and History of the Outer Planets: Theoretical Models and Observational Constraints," *Annual Review of Astronomy and Astrophysics* 22 (1984): 389.

Rasio, F. A., P. D. Nicholson, S. L. Shapiro, and S. A. Teukolsky. "Planetary System in PSR1257+12: A Crucial Test," *Nature* 355 (1992): 325.

Rasio, F. A., S. L. Shapiro, and S. A. Teukolsky. "Formation of a 'Planet' by Rapid Evaporation of a Pulsar's Companion," *Astronomy & Astrophysics* 256 (1992): L35.

Rebolo, R., E. L. Martin, G. Basri, G. W. Marcy, and M. R. Zapatero-Osorio. "Brown Dwarfs in the Pleiades Cluster Confirmed by the Lithium Test," *Astrophysical Journal* 469 (1996): L53.

Rebolo, R., E. L. Martin, and A. Magazzu. "Spectroscopy of a Brown Dwarf Candidate in the Alpha Persei Open Cluster," *Astrophysical Journal* 389 (1992): L83.

Rebolo, R., M. R. Zapatero-Osorio, and E. L. Martin. "Discovery of a Brown Dwarf in the Pleiades Star Cluster," *Nature* 377 (1995): 129.

Rees, M. "Opacity-limited Hierarchical Fragmentation and the Masses of Protostars," *Monthly Notices of the Royal Astronomical Society* 176 (1976): 483.

Reeves, H., ed. *Symposium on the Origin of the Solar System.* Paris: Centre National de la Recherche Scientifique, 1972.

Reuyl, D., and E. Holmberg. "On the Existence of a Third Component in the System 70 Ophiuchi," *Astrophysical Journal* 97 (1943): 41.

Richards, D. W., G. H. Pettengill, C. C. Counselman, and J. M. Rankin. "Quasi-sinusoidal Oscillation in Arrival Times of Pulses from NP 0532," *Astrophysical Journal* 160 (1970): L1.

Safronov, V. S. *Evolution of the Protoplanetary Cloud and Formation of the Earth and the Planets.* Moscow: Nauka, 1969.

Sagan, C. "Circumstellar Habitable Zones: An Introduction," in *Circumstellar Habitable Zones,* ed. L. R. Doyle. Menlo Park, Calif.: Travis House Publications, 1996.

Sagan, C. "So Many Suns, So Many Worlds," *Parade Magazine,* June 9, 1996, 11.

Sargent, A. I., and S. Beckwith. "Kinematics of the Circumstellar Gas of HL Tauri and R Monocerotis," *Astrophysical Journal* 323 (1987): 294.

Scherer, K., H. Fichtner, J. D. Anderson, and E. L. Lau. "A Pulsar, the Heliosphere, and Pioneer 10: Probable Mimicking of a Planet of PSR B1257+12 by Solar Rotation," *Science* 278 (1997): 1919.

Shao, M., and M. M. Colavita. "Long-baseline Optical and Infrared Stellar Interferometry," *Annual Review of Astronomy and Astrophysics* 30 (1992): 457.

Shklovskii, I. S., and C. Sagan. *Intelligent Life in the Universe.* New York: Dell, 1966.

Silk, J. "On the Fragmentation of Cosmic Gas Clouds. II. Opacity-limited Star Formation," *Astrophysical Journal* 214 (1977): 152.

Skrutskie, M. F., W. J. Forrest, and M. A. Shure. "Direct Infrared Imaging of VB8," *Astrophysical Journal* 312 (1987): L55.

Smith, B. "10 Years of Beta Pictoris—A Personal Reminiscence," in *Circumstellar Dust Disks and Planet Formation,* eds. R. Ferlat and A. Vidal-Madjar. Paris: Centre National de la Recherche Scientifique, 1994.

Smith, B. A., and R. J. Terrile. "A Circumstellar Disk around Beta Pictoris," *Science* 226 (1984): 1421.

Stauffer, J. R., D. Hamilton, and R. G. Probst. "A CCD-based Search for Very Low Mass Members of the Pleiades Cluster," *Astronomical Journal* 108 (1994): 155.

Stauffer, J. R., T. Herter, D. Hamilton, G. H. Rieke, M. J. Rieke, R. Probst, and W. Forrest. "Spectroscopy of Taurus Cloud Brown Dwarf Candidates," *Astrophysical Journal* 367 (1991): L23.

Stevenson, D. J. "Formation of the Giant Planets," *Planetary and Space Science* 30 (1982): 755.

Strand, K. A. "61 Cygni as a Triple System," *Astronomical Society of the Pacific* 55 (1943): 29.

Tavani, M., and L. Brookshaw. "The Origin of Planets Orbiting Millisecond Pulsars," *Nature* 356 (1992): 320.

Tombaugh, C. W. "Reminiscences of the Discovery of Pluto," *Sky & Telescope,* March 1960, 264.

van de Kamp, P. "Planetary Companions of Stars," in *Vistas in Astronomy,* ed. A. Beer. London: Pergamon Press, 1956.

van de Kamp, P. "Barnard's Star as an Astrometric Binary," *Sky & Telescope,* July 1963, 8.

van de Kamp, P. "Astrometric Study of Barnard's Star from Plates Taken with the 24-inch Sproul Refractor," *Astronomical Journal* 68 (1963): 515.

van de Kamp, P. "Parallax, Proper Motion, Acceleration, and Orbital Motion of Barnard's Star," *Astronomical Journal* 74 (1969): 238.

van de Kamp, P. "Alternate Dynamical Analysis of Barnard's Star," *Astronomical Journal* 74 (1969): 757.

van de Kamp, P. "Astrometric Study of Barnard's Star from Plates Taken with the Sproul 61-cm Refractor," *Astronomical Journal* 80 (1975): 658.

van de Kamp, P. *Dark Companions of Stars*. Dordrecht, The Netherlands: D. Reidel, 1986.

Walker, G. A. H. "On the Wings of Pegasus," *Nature* 378 (1995): 332.

Walker, G. A. H. "A Solar System Next Door," *Nature* 382 (1996): 23.

Walker, G. A. H. "One of Our Planets Is Missing," *Nature* 385 (1997): 775.

Walker, G. A. H., D. A. Bohlender, A. R. Walker, A. W. Irwin, S. L. S. Yang, and A. Larson. "Gamma Cephei: Rotation or Planetary Companion?" *Astrophysical Journal* 396 (1992): L91.

Walker, G. A. H., A. R. Walker, A. W. Irwin, A. M. Larson, S. L. S. Yang, and D. C. Richardson. "A Search for Jupiter-mass Companions to Nearby Stars," *Icarus* 116 (1995): 359.

Weidenschilling, S. J. "Aerodynamics of Solid Bodies in the Solar Nebula," *Monthly Notices of the Royal Astronomical Society* 180 (1977): 57.

Weidenschilling, S. J. "Formation Processes and Time Scales for Meteorite Parent Bodies," in *Meteorites and the Early Solar System*, eds. J. F. Kerridge and M. S. Matthews. Tucson: University of Arizona Press, 1988.

Wetherill, G. W. "Formation of the Earth," *Annual Review of Earth and Planetary Sciences* 18 (1990): 205.

Wetherill, G. W. "Occurrence of Earth-like Bodies in Planetary Systems," *Science* 253 (1991): 535.

Wetherill, G. W. "How Special Is Jupiter?" *Nature* 373 (1995): 470.

Wetherill, G. W. "The Formation and Habitability of Extra-solar Planets," *Icarus* 119 (1996): 219.

Wolszczan, A. "Confirmation of Earth-mass Planets Orbiting the Millisecond Pulsar PSR B1257+12," *Science* 264 (1994): 538.

Wolszczan, A., and D. A. Frail. "A Planetary System around the Millisecond Pulsar PSR1257+12," *Nature* 355 (1992): 145.

Zapatero-Osorio, M. R., E. L. Martin, and R. Rebolo. "Brown Dwarfs in the Pleiades Cluster. II. J, H, and K Photometry," *Astronomy & Astrophysics* 323 (1997): 105.

Zuckerman, B., and E. E. Becklin. "Excess Infrared Radiation from a White Dwarf—An Orbiting Brown Dwarf?" *Nature* 330 (1987): 138.

Zuckerman, B., T. Forveille, and J. H. Kastner. "Inhibition of Giant-planet Formation by Rapid Gas Depletion around Young Stars," *Nature* 373 (1995): 494.

INDEX